Rose Canker And Its Control

Paul Johnson Anderson

BULLETIN No. 183   MAY, 1918

MASSACHUSETTS
AGRICULTURAL EXPERIMENT STATION

Rose Canker and Its Control

By P. J. ANDERSON

This bulletin records results of investigations on a new
and serious fungous disease of roses, and describes
successful control methods

Recd, Sept. 23, 1918.

W. G. FARLOW

Requests for bulletins should be addressed to the
AGRICULTURAL EXPERIMENT STATION,
AMHERST, MASS.

Massachusetts Agricultural Experiment Station.

OFFICERS AND STAFF.

COMMITTEE.

STATION STAFF.

Administration.

WILLIAM P. BROOKS,[1] Ph.D., *Director.*
FRED W. MORSE, M.Sc., *Acting Director.*
JOSEPH B. LINDSEY, Ph.D., *Vice-Director.*
FRED C. KENNEY, *Treasurer.*
CHARLES R. GREEN, B.Agr., *Librarian.*
Mrs. LUCIA G. CHURCH, *Clerk.*
Miss F. ETHEL FELTON, A.B., *Clerk.*

Agricultural Economics.

ALEXANDER E. CANCE, Ph.D., *In Charge of Department.*
SAMUEL H. DeVAULT, A.M., *Assistant.*

Agriculture.

WILLIAM P. BROOKS,[1] Ph.D., *Agriculturist.*
HENRY J. FRANKLIN, Ph.D., *In Charge of Cranberry Investigations.*
EDWIN F. GASKILL, B.Sc., *Assistant Agriculturist.*
ROBERT L. COFFIN, *Assistant.*

Botany.

A. VINCENT OSMUN, M.Sc., *Botanist.*
GEORGE H. CHAPMAN, Ph.D., *Research Physiologist.*
PAUL J. ANDERSON, Ph.D., *Associate Plant Pathologist.*
ORTON L. CLARK, B.Sc., *Assistant Plant Physiologist.*
W. S. KROUT, M.A., *Field Pathologist.*
Mrs. S. W. WHEELER, B.Sc., *Curator.*
Miss ELLEN L. WELCH, A.B., *Clerk.*

Entomology.

HENRY T. FERNALD, Ph.D., *Entomologist.*
BURTON N. GATES, Ph.D., *Apiarist.*
ARTHUR I. BOURNE, A.B., *Assistant Entomologist.*
STUART C. VINAL, M.Sc., *Assistant Entomologist.*
Miss BRIDIE E. O'DONNELL, *Clerk.*

Horticulture.

FRANK A. WAUGH, M.Sc., *Horticulturist.*
FRED C. SEARS, M.Sc., *Pomologist.*
JACOB K. SHAW, Ph.D., *Research Pomologist.*
HAROLD F. TOMPSON, B.Sc., *Market Gardener.*
Miss ETHELYN STREETER, *Clerk.*

Meteorology.

JOHN E. OSTRANDER, A.M., C.E., *Meteorologist.*

---

[1] On leave.

Microbiology.

CHARLES E. MARSHALL, Ph.D., *In Charge of Department.*
ARAO ITANO, Ph.D., *Assistant Professor of Microbiology.*
GEORGE B. RAY, B.Sc., *Graduate Assistant.*
Miss LOUISE HOMPE, A.B., *Graduate Assistant.*
HAROLD L. SULLIVAN, B.Sc., *Graduate Assistant.*

Plant and Animal Chemistry.

JOSEPH B. LINDSEY, Ph.D., *Chemist.*
EDWARD B. HOLLAND, Ph.D., *Associate Chemist in Charge* (*Research Division*).
FRED W. MORSE, M.Sc., *Research Chemist.*
HENRI D. HASKINS, B.Sc., *Chemist in Charge* (*Fertilizer Division*).
PHILIP H. SMITH, M.Sc., *Chemist in Charge* (*Feed and Dairy Division*).
LEWELL S. WALKER, B.Sc., *Assistant Chemist.*
CARLETON P. JONES, M.Sc., *Assistant Chemist.*
CARLOS L. BEALS, M.Sc., *Assistant Chemist.*
WINDOM A. ALLEN, [1] B.Sc., *Assistant Chemist.*
JOHN B. SMITH, [1] B.Sc., *Assistant Chemist.*
ROBERT S. SCULL, [1] B.Sc., *Assistant Chemist.*
HAROLD B. PIERCE, B.Sc., *Assistant Chemist.*
JAMES T. HOWARD, *Inspector.*
HARRY L. ALLEN, *Assistant in Laboratory.*
JAMES R. ALCOCK, *Assistant in Animal Nutrition.*
Miss ALICE M. HOWARD, *Clerk.*
Miss REBECCA L. MELLOR, *Clerk.*

Poultry Husbandry.

JOHN C. GRAHAM, B.Sc., *In Charge of Department.*
HUBERT D. GOODALE, Ph.D., *Research Biologist.*
Miss GRACE MacMULLEN, B.A., *Clerk.*
Mrs. NETTIE A. GILMORE, *Clerk.*

Veterinary Science.

JAMES B. PAIGE, B.Sc., D.V.S., *Veterinarian.*
G. EDWARD GAGE, [1] Ph.D., *Associate Professor of Animal Pathology.*
JOHN B. LENTZ, [1] V.M.D., *Assistant.*

---

[1] On leave on account of military service.

CONTENTS.

PUBLICATION OF THIS DOCUMENT APPROVED BY THE SUPERVISOR OF ADMINISTRATION

BULLETIN No. 183.

DEPARTMENT OF BOTANY.

ROSE CANKER AND ITS CONTROL.[1]

BY P. J. ANDERSON.

INTRODUCTION.

Rose canker is a serious disease of greenhouse roses which was first described in 1917. It has probably been long prevalent in America, but has escaped notice largely on account of its obscure symptoms and consequent difficulty of diagnosis. Its ravages were formerly assigned to other causes or left unexplained. Rose growers who first brought it to the attention of this station in November, 1916, stated that they had been suffering severe losses for at least four years. After conditions in the rose houses had been investigated, the situation was considered so serious that work was immediately begun to determine more of the nature of the disease, and especially to find a remedy for it. The investigation was started in co-operation with L. M. Massey, pathologist of the American Rose Society, who first observed the disease two months before this, and had already decided that its seriousness warranted a thorough investigation. Research at the Massachusetts station has been largely confined to determination of the best methods of controlling the disease and investigation of such facts in the life history of the causal fungus as have a direct bearing on control measures. Massey undertook investigation of other phases of the disease, and has recently published his results (1917). A successful method of control has been evolved and is presented in this bulletin, but it is hoped that, as a result of long-term experiments now in progress in commercial houses, this method will be improved and, possibly, other easier methods found. However, since this will require a number of years, the present method is published in order that rose growers who are troubled with the disease may have the benefit of all that we already know about canker and its control.

---

[1] The writer is greatly indebted to Prof. A. Vincent Osmun, head of the department of botany at this station, for much valuable assistance, suggestions and criticism of the manuscript of this bulletin.

Only roses under glass are known to be affected. Some varieties, e.g. Hoosier Beauty, are more susceptible than others, but there is yet n evidence that any are immune. Massey (1917) observed the disease c Hoosier Beauty, Ophelia, Hadley, Russell, Sunburst, American Beaut and many seedlings. It has been reported only from the northern an eastern United States, but closer observation will probably show tha it has a much wider range.

SYMPTOMS.

The disease is most easily recognized by brown dead areas (cankers in the bark of the stems. These are more frequent and larger at th crown than higher up, but any part of the stem or branches may be at tacked. Crown cankers may be below the surface, just at the surface or, more often, extending up the stem, sometimes several inches (Plate I. Fig. 1). They may be confined to one side or may girdle the stem. Th young canker is blue-black or purplish in color and smooth, but as i becomes older the part above ground becomes reddish brown, dry, hare and cracked longitudinally. The margin is definite, and the dead area becomes sunken. Frequently the part of the stem immediately above the canker is swollen (Plate II.). When the subterranean part of the canker becomes old it is soaked and "punky," and the bark may be rubbed off between the thumb and forefinger, or it may rot away entirely (Plate I., Fig. 1). Sometimes a callus is formed around the edge of the canker.

Two types of cankers occur on the stem and branches higher up. The larger ones start from wounds, especially the stubs which are left after the blossoms are cut (Plate I., Fig. 2). Cankers from these stubs run back down the stems. The canker may stop at the first live branch below, but very commonly it continues to progress downward, and each successive branch dies as it is encircled by the descending canker. Cankers may also start from other wounds besides cut stubs. They are usually oval in outline and may be several inches long. The second type of aerial canker does not originate with wounds, but starts directly in the healthy green bark. First, small round purple areas appear, sometimes singly but more often in groups. As these increase in size the centers become light brown and the margins remain dark, giving a "bird's-eye" effect. When they occur in groups they coalesce and form large irregular dead areas in which, however, the individual cankers may still be distinguished for some time (Plate III., Fig. 2).

The depth of the canker varies, depending on such factors as the age of the part attacked, size of the infection court, environmental conditions and probably others. This is particularly a disease of the bark, and commonly the discolored area will be located outside the cambium entirely. But in more severe cankers it may extend to, or entirely through, the pith. If the shoot is young and has not yet hardened, the canker goes deeper and the entire shoot dies. This is frequently evidenced in the

PLATE I.

Fig. 1.—Old canker running up from the crown.

Fig. 2.—Canker running down from a cut stub.

PLATE III.

FIG. 1. — Canker resulting from coalescence of a number of small ones from stomatal infections.

FIG. 2. — Five cankers on a single stem.

sudden wilting and dying of shoots which have grown up rapidly from below the surface of the ground. Older shoots are rarely killed outright.

Only occasionally have we seen entire plants killed by this disease. One, several or all of the shoots of a plant may be attacked. Dead "brush" and dead small shoots are usually much in evidence in affected houses. The seriousness of the disease, however, lies not in the number of plants killed but in the fact that affected plants are small and weaker, resulting in diminished yields of inferior roses. The diseased plants cannot be forced, no matter how much fertilizer is applied and how well they are cultivated. New shoots do not grow from beneath the surface of the soil, but all come from the tops. These latter symptoms are the ones which the florist usually notices first, and, in fact, may be the only ones he notices.

Diagnosis of this disease is rendered difficult by two natural developments in the life of the rose plant which may easily be confused with disease: (1) Many varieties of roses naturally turn black at the crown very early; this, however, is a superficial blackening, and rarely runs up much above the surface of the ground. (2) The bark of all rose stems cracks with age, especially at the base, just as the bark of trees does. These two developments often resemble canker so closely that even one experienced in diagnosis may be misled.

DESCRIPTION OF THE CAUSAL FUNGUS.

Rose canker is produced by the parasitic growth of a fungus, *Cylindrocladium scoparium* Morg., within the tissues of the host (rose plant). Previous to 1917 this fungus had not been reported as a parasite. It was first found in Ohio by Morgan (1892) growing on an old pod of the honey locust (*Gleditsia triacanthus* L.). Seven years later it was reported again by Ellis and Everhart (1900) as growing on dead leaves of the papaw tree (*Asimina triloba* Dunal), and described as a new species, *Diplocladium cylindrosporum* E. and E.; but a study of the type materials of the two species by Massey showed them to be the same. As far as the literature shows, these are the only times that the organism had been observed up to 1916, and both times as a saprophyte.

The body of the fungus is composed of (1) mycelium, (2) sclerotia, (3) sporophores (conidiophores), and (4) spores (conidia). These four parts, or organs, of the fungus are here described separately.

Mycelium.

The mycelium is the part of the parasite which lives inside the tissues of the rose stem. It is composed of many microscopically slender, branching, tubular threads (hyphæ) which grow in every direction through the host cells for the purpose of securing nourishment from them for the fungus. Incidentally, in this process, the cells are killed and turn brown, thus producing the canker. The hyphæ are 4 to 6 *u* in diameter, and are divided by cross-walls (septa) into cells 5 to 20 times as long as their diameter. The manner of branching and septation is shown in Fig. 1. When the mycelium is young the walls are thin and not constricted, or, at most, only slightly constricted, at the septa. The contents consist of homogeneous protoplasm. Both the walls and contents are colorless,

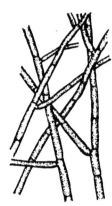

and when seen in mass, in pure culture, look like white cotton. But when the mycelium becomes older it becomes brown, the hyphæ are gnarled and twisted, deeply constricted at the septa, the cells short and oval or globose, giving one the impression of strings

Fig. 1.—Young mycelium from culture.

of beads (Fig. 2). The cells now contain large drops of

Fig. 2.—Old mycelium, showing chlamydospores.

reserve food, and the walls are thick. These cells are probably more resistant to adverse conditions, and serve to carry the fungus through unfavorable periods. They may be called chlamydospores. Their diameter is much greater than that of the ordinary hyphæ, as indicated by the figures.

Sclerotia.

Sometimes the surface of old cankers is dotted over with minute shining black pimples (Plate II.). They are usually not much larger than a pin

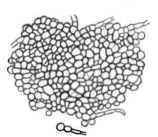

Fig. 3.—Thin section through a sclerotium.

point and never as large as a pin head. To the naked eye they look like pycnidia, but microscopic examination always proves them to be sterile balls of thick-walled pseudoparenchymatous fungous cells (typical sclerotia). They are directly under the epidermis, but this does not obscure their shining black prominence. In certain culture media they are produced in great abundance. The cells are much like the chlamydospores; in fact, the sclerotia seem to be only a further development of the chlamydospore-forming hyphæ, and all gradations between the two may be found. Their function is probably the same as that of the chlamydospores. A thin cross-section of one is shown in Fig. 3.

CONIDIOPHORES.

The conidia, or ordinary spores, — as distinguished from the chlamydospores, — are borne on special upright branches, — conidiophores. These are produced in great abundance in artificial culture, but are rarely seen on the cankers. The writer has found them occasionally just at the surface of the ground on young shoots recently killed by the pathogene. But in badly infested rose beds which are kept wet they are produced in great abundance on dead shoots and parts of the rose plants which are cut off and left to decay on the ground under the bushes. To the naked eye the dead shoots seem to be dusted over in patches with a white powder. Under a strong hand lens — or better, a binocular microscope — each particle of this white powder is seen to be composed of a tuft of

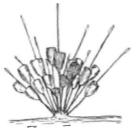

Fig. 4. — Tuft of conidiophores on a dead rose stem.

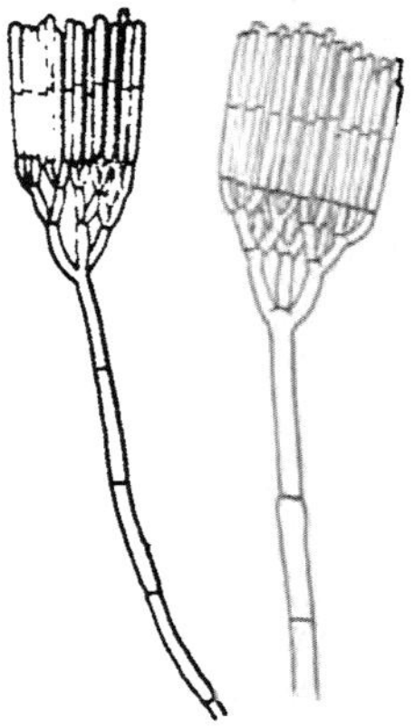

slender-stalked "brooms" with glistening white heads. One of these tufts is shown in Fig. 4. Each little broom is a conidiophore with its mass of conidia on the apex. The number of conidiophores in a tuft varies from 5 to 40, or more. No details, further than shown by Fig. 4, can be made out under the binoculars. Under the compound microscope, however, it is possible to determine accurately the structure of these little brooms. Examined in the dry condition they appear as in Fig. 5, where the conidia are cemented to-

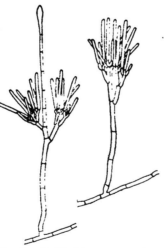

Fig. 5.—Conidiophores and conidia as seen in a dry condition.

Fig. 6. — Conidiophores as seen when mounted in water, many of the conidia washed away.

gether into a solid head. But when mounted in water the cement which holds them together dissolves, many of them float away, and the head becomes loose as represented in Fig. 6. The main stem of the conidiophore may be unbranched up to just below the conidia, as represented by Fig. 5, or it may show one or more monopodial branches at various heights.   The spores are frequently borne on lateral branches of this stem (Fig. 6), while the main stem is continued upward and terminates in an enlarged club.   The ultimate branchlets, and one or two series below them, are usually in threes, as shown in Fig. 5, but twos are not uncommon.   In regard to the dimensions of the conidiophore, Morgan (1892) writes: "the fertile hyphæ have a simple septate stem 5 to 7 μ in thickness, and are dissolved above into a level-topped cyme of branches; their height, exclusive of the spores which easily fall off, is 125 to 150 μ."   Ellis and Everhart (1900) give the dimensions as 50–110 x 5–6 μ.   In pure culture the writer has found them taller than the above measurements; an average of 50 conidiophores grown on potato agar gave 291 μ, and the diameter of the stalk, 6.6. μ.

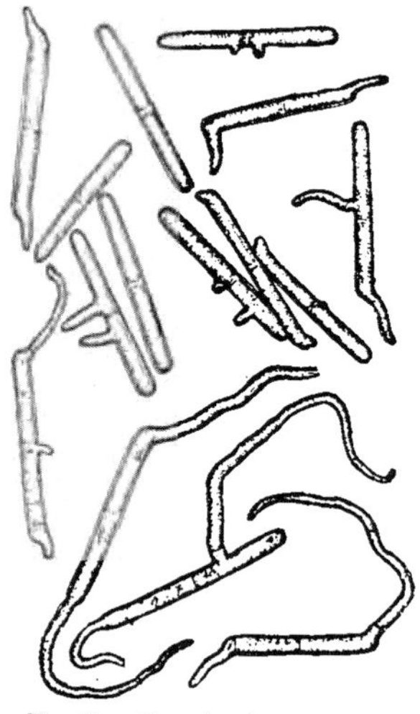

CONIDIA.

The conidia are long, cylindrical, obtuse at each end, hyaline, divided into 2 cells by a septum at the center (Fig. 7). The contents are at first homogeneous, but later show vacuoles or oil drops (Fig. 8). Morgan

FIG. 7. — Germinating conidia.

FIG. 8. — Old conidia.

(1892) gives the dimensions as 40–50 x 4 μ at the apex, and 3 μ at the base; Ellis and Everhart (1900), 40–50 x 4–5 μ; Massey (1917), 36–55 x 3.3–4.51 μ, with an average of 48.3 x 4.13 μ. The writer found the average of 50 on a young potato agar culture to be 48.8 x 5.1 μ; 50 on a two-months' culture, 39.2 x 4.03 μ; 50 produced on a pod of Gleditsia, 41 x 4.1 μ.

LIFE HISTORY OF THE FUNGUS.

Before any measure of control could be intelligently attempted it was first necessary to become intimately acquainted with the life history of the causal organism (the pathogene). In the studies which are recorded below most attention was directed to those points which appeared to have a direct connection with control. Nevertheless, in order to become familiar with the entire life cycle, certain phases of development which have no obvious connection had to be investigated. For convenience in discussion, the life history is treated under three heads: —

1. Germination of the spores.
2. Parasitic life of the fungus (pathogenesis).
3. Saprophytic life of the fungus.

Germination of the Spores.

The life cycle begins with germination of the spores. The first essential condition for germination is the presence of water. Spores never germinate except when they are directly in water. A moist atmosphere is not sufficient. Germination takes place through the production of one or more tubes from each of the two cells of the spore. Usually the tubes do not start at the same time; one in each cell begins to grow, and this is later followed by another. Four germ tubes to each spore is the most frequent condition, but there may be more or fewer. The tubes may come out from any place on the surface of the spores, as illustrated in Fig. 7. They elongate very rapidly at laboratory temperatures, quickly develop septa, branch repeatedly and soon a mycelium is produced.

The brown thick-walled cells of the mycelium, which we have called chlamydospores, germinate by the production of slender hyaline germ tubes similar to those of the conidia and under the same conditions. Other detached cells of the mycelium also possess the power of germination. Especially is it common to see germ tubes arising from the cells of the main stem of the conidiophore when detached and kept in water. Such germ tubes usually arise from the end walls of the cells, and may grow directly through one or more old cells before emerging.

Temperature Relations.

The relation of temperature to germination of spores was studied carefully in the hope of evolving some method of control by keeping the rose houses at temperatures which are unfavorable for germination and thus retarding progress of the disease. The general effect of variation of temperature and the maximum, minimum and optimum temperature for germination were determined by the following method: —

Method. — Viable spores from a young, pure culture were transferred to a drop of water in the center of a glass slide. The slide was supported on two short glass rods in a Petri dish, used as a moist chamber. A few drops of water placed in the bottom of the dish kept the air humid and prevented drying out of the drop containing the spores. The Petri dish was then kept at the desired constant temperature in incubator, refrigerator or constant temperature room. Observations were taken and percentages of germination counted at regular intervals. No figures are based on the results from a single slide. Each result tabulated represents the average of several slides. Tests at high or low temperatures were controlled by duplicates at ordinary room temperatures.

The results of the tests are summarized in Table I.

TABLE I. — *Effect of Temperature Variation on Spore Germination.*

Temperature, Centigrade (Degrees).	Period before starting to germinate (Hours).	Percentage of Germination in 24 Hours.
5, ...	–	0
8–9, ...	24	1 (2 per cent. in 48 hours).
12, ...	5	95
15, ...	Not observed before 7 hours, when about 20 per cent. had started.	95
17, ...	4–5	95
20, ...	4½	95
22–23, ...	3–4	95
25–26, ...	2–3½	95
28, ...	Not observed before 6½ hours, when 95 per cent. had germinated.	95
30, ...	6½	95
31, ...	6½	70
33.5, ...	6½	21 (Erratic and abnormal).
36, ...	4	70
37.5, ...	–	0
40, ...	–	0

It is apparent from these tests that spores germinate at any temperature between 8° and 36° C. Between 12° and 30° the percentage of germination was almost total, ranging from 95 to 100 per cent. (all marked 95 per cent. in the table). Within these limits there was practically no variation of percentage due to temperature. In other words, if the optimum temperature is to be determined by percentage germination alone, it is very wide. Below 12° the percentage drops off rapidly until at 8° to 9° we get but 1 per cent. in twenty-four hours. Germination ceases altogether below this. Between the temperatures of 31° and 36° it is difficult to express the effects of temperature in percentages. Not only is germination erratic, varying greatly in slides apparently treated alike, but it may also be so abnormal that it is difficult to determine just what constitutes germination. The spores assume peculiar shapes by the development of knobs or, more commonly, globose swellings twice the diameter of the spores. These vary in number and location, but most frequently they are on the ends of the spores. Very slender unbranched germ tubes may grow for a time from these. The percentage of spores affected does not gradually diminish to form a regular curve. Thus, in one test at 36°, 70 per cent. were affected in this way. But at 37.5° there was no germination or change in the spores which could be detected with the microscope. The effect of temperature variation is more apparent in the *time required* for germination to begin than in the final *percentage* of germination. In this respect there is a rather regular curve. The optimum is at about 25°, where germination begins in two to three and one-half hours. At 12° it required five hours, and at 8° no germination was apparent until after twenty-four hours. The fact that spores do not germinate at a certain temperature does not mean that they are dead. Spores kept for two days at 5° showed not the least indication of germination, but when brought back to ordinary room temperatures· they quickly germinated to over 95 per cent. Experiments to be described later show that spores may be kept for long periods at temperatures both lower and higher than indicated in this table and still retain their viability.

Apparently there is little opportunity for retarding the progress of the disease by maintaining temperatures in the house unfavorable to the fungus, because the optimum temperature for spore germination is approximately the same as the optimum for growing roses. The latitude of the germination optimum is also unfavorable to such a method of control.

Effect of freezing the Spores.

It is a well-known fact that the spores — especially the conidia — of many fungi are quickly killed by freezing, and this weakness may be utilized in checking disease. The purpose of the present investigation was to determine whether the spores of Cylindrocladium can be killed by freezing, and if so, how much exposure is required. Two methods were used.

First Method. — Petri dishes containing young cultures with abundance of spores were exposed to out-of-door temperatures of —3° to —10°C. Checks were first made at room temperatures to test the viability of the spores. Spores were removed from the frozen plates at regular intervals and put to germinate in moist chambers at ordinary room temperatures, as described above in spore germination tests. By this method the spores were dry when frozen.

After about two hours the percentage of germination began to decline; in eight hours it had fallen to 10 per cent.; in twelve hours, to less than 1 per cent.; and at the end of fourteen hours there was no germination whatever. All checks germinated 95 per cent.

Second Method. — Spores were transferred from plates along with a portion of the agar to drops of water on slides. All was macerated until the spores were well distributed through the water. They were immediately put outside to freeze and one slide brought into the laboratory at the end of each hour and tested for germination.

The results were very similar to those obtained by the first method. Freezing for one hour seemed not to affect them at all; in two hours the percentage dropped to from 75 to 80 per cent.; in three hours, to 30 per cent.; in six and one half hours, to 25 per cent.; in ten hours, to 1 per cent. From 1 to 2 per cent. germinated even after exposures of twenty-four hours, but these were spores in the center of the drop of water, or directly in the agar, which seemed to give them some protection. There was no germination whatever after thirty-six hours.

The first method more nearly approximates natural conditions, but under any conditions we may safely draw the conclusion from these experiments that all spores are killed by freezing during thirty-six hours.

Thermal Death Point of Spores.

Investigation of this point was undertaken with a view to the possibility of sterilization by heat. Thermal death point is defined as the lowest temperature at which an organism is killed by an exposure for ten minutes. Since this point might be different for spores than for mycelium, each was tried separately.

Method. — Spores from a young culture immersed in a drop of water were placed in a thin pipette tube, sealed at one end and covered with a rubber cap at the other. The tubes were then dropped into vessels of water kept at the desired temperature. Each vessel was supplied with a thermometer, and could be heated by a Bunsen burner when necessary. After ten minutes the tubes were removed, the sealed end filed off, and the spores forced out through it on to a glass slide by pressing the rubber cap at the other end. The slides were then put in moist chambers as previously described in germination tests. These were kept at ordinary laboratory temperatures. Temperatures at intervals of 1°, from 40° to 55°, were tried. All tests were made in duplicate several times.

Up to and including 46° the spores did not seem to be affected by ten-minute exposures. Above this the percentage remaining alive declined very rapidly to the absolute thermal death point of 49°. At this temperature none ever germinated.

It was also found that spores can be killed at lower temperatures than 49° by exposing them for longer periods. In some previous experiments it had been determined that they are killed by an exposure to 37.5° for twenty-four hours. At 42° they are killed in two hours. To determine the effect of varying the period of exposure at a given temperature, 40° was selected as a standard, and spores exposed (in drops of water on slides in Petri dishes) during periods differing by intervals of one hour. They were then brought back to room temperature and tested as above. The results of this series are given in Table II.

TABLE II. — *Germination of Spores after Exposure to a Temperature of 40° C.*

PERIOD OF EXPOSURE (HOURS).	Time required after Removal to Room Temperature before beginning to germinate.	Percentage of Germination after 24 Hours at Room Temperature (20–24° C).
1,	95 per cent. in 3½ hours. Not observed sooner.	Over 95
2,	2½ hours. Just starting.	Over 95
3,	3 hours. Just starting.	Over 95
4,	60 per cent. after 5 hours.	Over 95
5,	At least longer than 4 hours.	50
5½,	1 per cent. in 7 hours.	2
6½,	At least longer than 6 hours.	0.5
7½,	– –	0
9, 12, 14, 18, 20,	– –	0

It will be noticed that the longer the period of exposure, the longer the time required for germination after being removed to room temperature. There was no decrease in the percentage of germination until after four hours. From this point it dropped rapidly to less than 1 per cent. in six and one-half hours, and no germination whatever after seven and one-half hours.

Effect of Desiccation on the Spores.

The length of time during which spores are able to live in a dry condition may have an important bearing on dissemination of a fungus and spread of a disease. Neither the thinness of the walls nor character of the spore contents of Cylindrocladium would lead one to expect great longevity. The following method was used to determine longevity at ordinary room humidity: —

Method. — The lids of Petri dishes, containing pure cultures of Cylindrocladium with abundance of conidia, were lifted enough to allow the thin film of agar to become hard and dry within a day or two. At intervals of one day spores were transferred from these dishes to drops of water on slides in Petri dishes, as previously described for other germination tests. The percentages of germination were determined after the spores were kept in moist chambers for twenty-four hours. All checks — made from the cultures before tilting the lids — germinated to over 95 per cent. Several hundred spores were transferred for each test. Three different Petri dish cultures were used at different times.

In every trial the percentage of germination began to decline after twenty-four hours. In two days it had dropped to 25 per cent.; in five days, to 10 per cent. After ten days not more than 1 per cent. germinated, and in no case was any germination observed after drying for fifteen days.

The longevity of conidia, then, appears to be very limited when kept in a dry condition. When the atmosphere is kept very humid they live longer, at least several weeks, but no careful investigation has been undertaken to determine just how long with each degree of humidity. If water stands on them, even in the culture dish, they germinate and then quickly die if dried out at once.

PARASITIC LIFE OF THE FUNGUS.

Pathogenicity.

In order to prove that an organism is the causal factor of a certain disease there are four requirements — called the four rules of proof — which pathologists all agree must be fulfilled. These are: (1) find the organism constantly associated with the disease; (2) isolate the organism, grow and study it in pure cultures; (3) produce the disease again by inoculation from these pure cultures; (4) reisolate the organism and prove by culture its identity with the organism which was first found. These four rules were complied with by Massey (1917), and the pathogenicity of *Cylindrocladium scoparium* established. The present writer has also given the four rules repeated test, and obtained results similar to those of Massey. These experiments are not described in detail here, but only certain notes on each of the four steps recorded.

1. Constant association of the pathogene with the canker is not so easy to establish as in most fungous diseases because the fungus can rarely be seen with the naked eye on cankers in rose houses. Nevertheless, the writer has occasionally been able to find a white band of conidia around cankers on young shoots just at the surface of the ground. Almost always when a canker is kept in a moist chamber for twenty-four hours or longer the mycelium grows out as long, straight, white hyphæ, which can readily be recognized as peculiar to Cylindrocladium by one who has become acquainted with the appearance of this fungus. Also, after a few days in the moist chamber, conidia usually begin to develop on the surface. The presence of the pathogene in old cankers is also often betrayed by sclerotia, — small, flat, shining black specks just under the epidermis. Yet the writer has often found cankers in which the organism could not be determined in any of the above ways. There seems to be only one absolutely sure way of determining association of the pathogene in all cases, and that is by making isolations, which is really a part of the second rule of proof.

2. The following has been found the most satisfactory method of isolation: —

Method. — The surface of the canker is first sponged with mercuric chloride 1–1,000. Scalpels and steel needles are kept in a jar of 95 per cent. alcohol. The epidermis, or at least a thin outer layer of the canker, is then peeled off with a scalpel from which the alcohol has been burned over a Bunsen. Another scalpel sterilized in the same way is used to cut out a portion of the peeled canker. It is then removed with a flamed needle to a flask of sterile water, washed, and transferred to a potato agar slant — or sometimes poured plates are used. One or two drops of lactic acid are added to the tube of agar when slanted. The acid not only prevents growth of bacteria, but also seems to make the medium more favorable for the growth of Cylindrocladium. Occasionally other agars, such as corn meal, oat, lima bean, Czapek's and Cook's No. 2, have been successfully used, and there is no objection to them. The almost constant use of potato agar in the present investigation is due more to habit and convenience than to any advantage over other media. In the case of small initial cankers the epidermis was not peeled off. The mycelium grows up into the air and into the agar very quickly, and after some experience one is able with the naked eye to distinguish within twenty-four hours the growth of Cylindrocladium from that of other fungi he is apt to meet with on roses. But if there is any doubt, he has but to wait another day or two, and spores are produced by which this fungus can be absolutely identified.

Other methods of isolation besides tissue transfers have been successfully used. Where spores are present, or where they have been developed in moist chambers, cultures are very easily made by touching them with the tip of a sterile platinum needle, — first thrusting the needle into the agar so that more spores will adhere, — and then transferring to agar slants. When the sclerotia were first discovered on the cankers there was some question as to their connection with Cylindrocladium. Some of them were picked out under the binoculars with a sterile needle, freed from all clinging rose tissue, washed in sterile water, and transferred to agar plates. In this way, also, pure cultures were obtained.

By the first method described, the organism has been isolated in pure culture from hundreds of typical cankers. In order to determine the very youngest stages, a number of stems showing the little round lesions (described under "Symptoms"), from the size of a pin point to several millimeters in diameter, were brought into the laboratory, washed merely with sterile water, and transfers made as above. Pure cultures were obtained from even the smallest of them.

The relation of the pathogene to dead stubs was also determined in this way. After the flower is cut, one or more shoots quickly grow out from below the cut end of the stem. The topmost one, however, is usually some distance below the cut surface, and a useless stub is left from 1 inch to 3 or 4 inches long. This stub usually dies slowly from the apex back to the first branch, where it is apt to stop. When the canker disease is prevalent in the house, however, the dying frequently does not stop at the first shoot but continues down the stem, and the shoots die as they are encircled by the descending dead area. Frequently the fruiting bodies of various species of fungi, such as Pestalozzia, Phoma, etc., can be found on these stubs, but in other cases no spores could be found. A large number of them were collected from a house known to be infested, and transfers made. Cylindrocladium was obtained from over half of them. After they were found to be infested in some cases, more attention was directed to them and the sclerotia frequently observed. It was from these sclerotia that the pure cultures mentioned above were obtained.

Study of the fungus in pure culture will be described later.

3. Plants were inoculated in four different ways: —

Methods of Inoculation. — (a) Stems wounded, inoculated with agar in which the fungus was growing, kept moist several days with moist cotton. (b) Same as (a), but the plants not wounded. (c) Wounded, spores sprayed over the plants with an atomizer, and kept for several days under a bell jar. (d) Same as (c), but plants not wounded. All these methods were controlled by checks treated in the same way except for applying the fungus.

Typical cankers were produced by all four methods of inoculation. The shortest incubation period — time between inoculation and first appearance of symptoms — was four days on the wounded plants and five days on the unwounded ones. The rate of development of the canker after it first appears varies greatly. On some plants which were first wounded and kept under bell jars the cankers were over a centimeter across in two weeks, but if the bell jars were removed and the humidity of the air diminished, the cankers grew very slowly. Small aerial cankers usually soon stop growing altogether unless several of them occur close together, or unless they are kept very moist. Crown cankers grow more rapidly than cankers higher up, but their rate of growth becomes decidedly slower as they advance above the surface of the soil.

4. Reisolations were very readily made from a number of the cankers produced by artificial inoculation. The fungus was obtained in pure culture, and easily identified by its cultural and morphological characters as *Cylindrocladium scoparium.*

Infection Court.

The artificial inoculations described above indicate that a wound is not necessary for infection. All observations indicate, however, that a wound is a very favorable infection court. A great many of the basal cankers start from the union of stock and scion; aerial cankers from the cut surfaces of stubs and from various bruises made by tools, etc. Even where no wound appeared, it seemed possible that there might be small wounds not readily visible to the naked eye. In order to determine whether such was the case, and if not, to determine whether any natural openings in the epidermis serve as infection courts, artificial inoculations were made by spraying spores with an atomizer on what, as far as could be seen with the naked eye, seemed to be perfectly healthy stems. As soon as cankers began to appear they were cut out, fixed, imbedded in paraffin, cut into serial sections and stained. Twenty-four cankers varying from the size of a pin point to 2 millimeters in diameter were used and cut serially to a thickness of 8 μ. In no case was any wound through the epidermis discovered. But in every case a stomate was located directly at or very near the center of the canker. In the larger cankers there were several stomates, and it was not always possible to determine the point of entry. In the smaller ones, however, only one was present, and it was always approximately at the center. A number of infections were also discovered which were so small that they had not been seen when the material was fixed. In some cases the affected cells extended no farther than 5 or 6 rows below the stomate.

There does not seem to be any reasonable doubt that the stomates serve as infection courts, and that the little round lesions on the smooth stems are largely the result of these stomatal infections.

The Mycelium in the Host Tissues.

In order to follow the course of the mycelium after it has entered the rose stem, and to determine its effect on the host tissues, cankers in every stage of development, from that where they are not yet visible to the naked eye up to the old, fully developed lesion, were sectioned, stained and studied.

Method. — The mycelium is very difficult to follow in unstained sections, but after some experimenting a simple method of treatment was found by which the mycelium could be very distinctly differentiated in the host cells. Cankers were fixed in Gilson's fluid, dehydrated gradually, and cut with a slide microtome from 95 per cent. alcohol. [1] The sections were then stained one minute in a saturated solution of safranin in 95 per cent. alcohol, excess safranin removed by transferring to 95 per cent. alcohol for one minute, stained one minute in 1 per cent. gentian violet in clove oil, and cleared in clove oil, the oil washed out with xylol and the sections mounted in balsam. This method is very rapid and any number of sections can be stained at one time.

Before describing the behavior of the mycelium in the tissues it will first be necessary to review briefly the structure of a normal rose stem. Fig. 9 represents a cross-section of a stem of about the age when cankers are most frequent.

Normal Structure of the Stem. — On cutting through a rose stem with a knife, one very readily notices that it is composed of three distinct parts, (1) a rather succulent outer cylinder of bark, (2) a central soft white pith, and (3) a hard cylinder of wood between the two. The cell elements which occur in each of these will be enumerated in order, beginning with the outside.

First, the stem is covered with a smooth, thin, waterproof coat, — the cuticle. Just beneath this is the one layer of rather flat cells composing the epidermis. Next in order are three or four layers of cells with heavy walls and no intercellular spaces. This is the collenchyma. The cuticle, epidermis and collenchyma form an air-tight, water-tight covering of the stem, uninterrupted except by the stomates. These microscopic breathing pores, which are not so numerous on the stem as in the leaves, are guarded and strengthened on either side by crescent-shaped projecting cells. The structure of the stomate can best be understood by reference to the figure. It will be noticed that there is a free passage between the guard cells into the stomatal cavity beneath, and from here to the loose, thin-walled cells of the next underlying tissue, the chloren-

---

[1] Very small cankers were imbedded in paraffin, sectioned and stained in the usual way; but for larger cankers this was found to be unnecessary, and a long and tedious process.

chyma. Except under the stomates, where it is thicker, the chlorenchyma is composed of three or four layers of cells containing around the inside of the walls the green chloroplasts which give the color to the bark. Next in order are the large thin-walled cells of the inner cortex, the lowermost

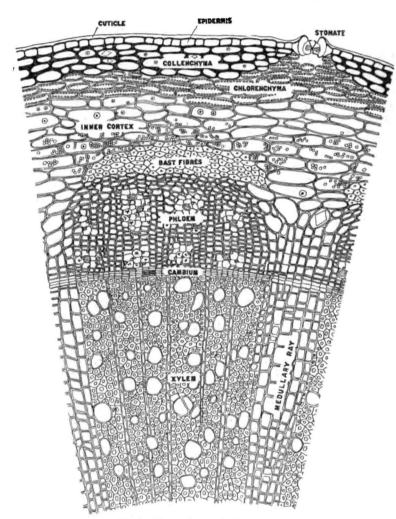

Fig. 9. — Transection of a healthy rose stem.

of which contain abundant starch grains in storage. Next there are areas of angular, very thick-walled cells, the bast fibers. The walls are so thick that there is hardly any opening (lumen) through the center. In longisection these are seen to be shaped like long, sharp-pointed pencils, with the sharp ends overlapping. Their function is to give rigidity and strength. The areas of bast fibers do not form a complete cylinder, but the inner cortex tissue runs down between them. Just under each bast area there is a region of tissue called phloëm. It contains long tubes (sieve tubes) through which the elaborated plant food passes down through the stem from the leaves. Each sieve tube is accompanied by a line of small slender cells (companion cells), which appear in transection as though they' were cut out of the corners of the sieve tubes. The remaining cells of the phloëm are box-like cells called phloëm parenchyma. The phloëm is bounded below by the cylinder of thin flat cells, the cambium, which marks the line of cleavage between the bark and wood.

The wood, or xylem, is composed mostly of four kinds of cells: (1) Box-like parenchyma cells which compose the broad medullary rays as well as the narrow rays one cell in width. (2) Long tubes of large diameter (tracheæ) through which the water mainly passes from the roots to the parts above. The walls are strengthened by spiral or annular thickenings. (3) Vertically elongated cells (tracheids) of smaller diameter and thicker walls, also water carriers. These make up the greater portion of the wood. (4) Wood fibers, somewhat smaller in diameter, with thick walls and long tapering points. They cannot be distinguished from the tracheids in transection. Although the walls of all the xylem elements are heavy, they are all marked with pits so that liquids have only a thin membrane through which they must pass to go from one cell to the next.

The pith (not shown in the figure) is composed of cells of only one kind, large or small, somewhat isodiametric (parenchyma). The walls are very thin.

Path of the Mycelium. — The germ tube, when it attacks the host, is very slender and easily passes between the guard cells down into the stomatal cavity. It could then readily pass between the loose cells of the chlorenchyma and inner cortex, but it does not choose to progress this way. Only rarely has the mycelium been seen progressing for any considerable distance between the cells, but it immediately passes *into* the cells by means of holes which it is able to dissolve through the walls. From this time on the mycelium is entirely intracellular except for the short distances through which it sometimes passes from one cell to another. It branches profusely, but the host cells do not become filled with mycelium. Rarely are more than one or two strands seen in a single cell, except in very old cankers. It is very slender and delicate at first, but in age becomes brown and takes on the various cell forms previously described for the mycelium. It seems to prefer the starch storage cells of the inner cortex, and in cankers of medium age is

FIG. 10. — Young mycelium in the cells of the inner cortex.

always found most abundantly in these cells (Fig. 10). However, the other cells are not immune. Mycelium may be found quite abundantly in the collenchyma, the heavy walls of which seem to offer no resistance whatever to the progress of the invader. Occasionally it has been found even in the epidermal cells. The first bar to its inward progress is the area of bast fibers. It does not pass through these at once, but in very old cankers it has been observed even in the bast fibers. There is, however, an easy path between the bast areas through the flaring outer ends of the medullary rays, which do not stop at the cambium but extend up between the phloëm areas. From here the hyphæ can easily pass laterally into the phloëm. Passing down into the xylem elements the invader finds its progress made much easier by the presence of pits in all the walls. It does not confine itself to the medullary rays, but passes laterally into the other elements. The mycelium has been found in every element of the xylem, least of all, however, in the wood fibers. Often in old cankers the tracheæ may be found almost clogged with mycelium, frequently in the form of chlamydospores. The method by which it passes through the walls is shown in Fig. 11. From the xylem it passes down into the pith, where it finds progress easy through the thin walls.

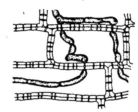

Fig. 11. — Mycelium in the cells of the medullary rays.

Effect on the Host Cells. — All of the cankers do not extend to the pith. A great many of them, for some unexplained reason, never go deeper than the bark. The fact that the affected plants stop growing, and do not send up any more shoots from below the cankers, is probably due to destruction of the phloëm, which prevents any food passing down to the lower stem or roots. The cells somewhat in advance of the invading hyphæ first become filled with a brown, finely granular substance which gradually becomes coarser and later mostly disappears, possibly being used by the parasite, and the cells are left almost empty. The starch, nuclei and chloroplasts also disappear. The walls are not affected except for the holes through which the hyphæ pass. The whole effect on the host seems to be entire disorganization of the cell contents. There is no hyperplasia, hypertrophy or other abnormal cell change in the canker. To be sure, there is often a swelling just *above* the canker, which is produced by an increase both in the size and number of cells of the inner cortex. This is, however, probably due to the amount of elaborated food which is stopped here because it cannot now continue downward on its normal course. As the canker becomes older, the cells of the bark collapse, being now empty. The cracks which then appear in the bark may be due to the contraction of the dying tissue, or to the expansion of the growing stem, or both. The cells of the xylem and pith do not collapse, but the affected tissues turn brown.

Saprophytic Life of the Fungus.

Early in this investigation it was discovered that the canker pathogene does not necessarily live all the time on the rose plant, but that it is also a natural inhabitant of the soil. This was first proved by isolating it under sterile conditions from soil 4 and 5 inches below the surface in the rose beds. Then it was found that when sterilized soil is inoculated the mycelium spreads rapidly through it and lives and grows normally there for a long time. Since these pure cultures in soil have been used rather extensively in this investigation, the method of making them is described here and omitted in all future references.

Method. — Milk bottles of 1 quart capacity were used. Thirty-three cubic inches of rose soil, moistened until muddy, was put in each bottle. The mouth of the bottle was then plugged with cotton and the whole sterilized in an autoclave. After it was cool it was inoculated by transferring a small bit of agar containing mycelium to the surface of the soil. Soil so treated becomes entirely infested in twelve to twenty-one days at ordinary room temperature.

Longevity of Mycelium in the Soil.

Before undertaking control measures it was very essential to know whether the fungus lives indefinitely in the soil, or whether it starves out and dies when the rose plant is not present to furnish nourishment. On March 27, 1917, eight milk bottles of soil were inoculated. At the end of every month clods of soil were transferred from these bottles to acidified agar plates. It has been found that when soil particles containing living mycelium are transferred to agar plates the mycelium begins to grow out on to the agar within twenty-four hours, and in a few days produces spores by which it can be definitely identified. The soil bottles were kept in a dry culture room. No water was added to them, but the soil is still somewhat moist at this writing. One year from the date of inoculation every plate isolation gave pure cultures of Cylindrocladium. There seems to be no doubt, then, that it will live for a year at least, and probably indefinitely, in the soil without the rose plant being present.

Growth on Other Substrata.

The longevity of the mycelium may possibly be increased by passing a part of its existence on substrata other than the living rose plant and the soil. The abundant growth and production of spores on dead and decaying rose twigs on the soil has previously been referred to. Dead rose leaves were sterilized and inoculated with spores in moist chambers, and it was found that the mycelium grows luxuriantly and produces some spores on them. Pods of the honey locust and leaves of the papaw tree — substrata on which the fungus was previously reported — were inoculated in the same way. The fungus grew normally on both, producing spores in great abundance on the pods, and less abundantly on the leaves. The great variety of artificial media on which it can be made to grow in the laboratory also indicates a wide range in feeding habits. Other kinds of decaying vegetable matter in the soil were not tried, but it would not be surprising if it were found capable of living on a great number of them.

Depth of Penetration of the Soil.

In the soil isolation tests the fungus was not found below 5 inches, but this was not conclusive, since the method of isolation proved not to be entirely satisfactory, and only a few isolations were made. The soil in the milk bottles was never more than 4 inches deep, but the fungus grew as luxuriantly at that depth as at the surface of the ground. In order to test its ability to penetrate to greater depth, glazed drain tiles 2 feet long were closed at the bottom with an inch of cement, filled with soil, plugged with cotton at the top and sterilized. The soil was then inoculated on the surface. Holes had been drilled at regular intervals through the side of the tiles. These were corked, and after the whole was sterilized the corks were made air-tight and water-tight by covering them with melted paraffin. In order to determine whether the fungus had penetrated to a certain depth a cork at that depth was removed, a portion of the soil next to it transferred to an agar plate, and the hole immediately made tight again, all operations being carried out under aseptic conditions. Unfortunately the soil became dry too quickly, due to the large opening at the top, and it was found necessary to pour more water on to the top of the soil. At this writing the fungus is growing throughout the entire depth of soil in the tiles, and has been isolated from the lowest holes, almost 2 feet below the surface. Whether it was washed down by the water or grew down naturally is not certain, but at present the fungus is growing normally in every particle of soil 2 feet below the surface. If it could be washed down by the water in the tiles, there is no reason why it should not be washed down by water in the rose houses. Judging from these results, and what is known about the penetration of other soil fungi, there seems to be no reason for doubting that the mycelium may exist several feet below the surface, depending to some extent on the character of the soil.

Rate of Growth of the Mycelium.

The rapidity with which mycelium grows through soil is dependent on the temperature. The optimum, maximum and minimum temperatures for growth were determined for the purpose of finding which temperatures in the greenhouse are favorable and which unfavorable to the spread of the fungus.

Method. — When the milk bottles of infested soil are kept in a dark place the progress of the white mycelium downward can be readily observed through the sides of the bottles. A number of bottles were inoculated, and when the mycelium was well started downward the limit was marked accurately by blue pencil lines around the bottles. The bottles were then placed simultaneously in incubators, ice boxes and constant temperature rooms, wherever a constant temperature could be maintained for a week at a time. A new line was drawn at the end of every forty-eight hours.

The results of this test are tabulated in Table III. An examination of this table shows that the optimum temperature for growth is 26 to 27° C, the minimum is just above 8.5°, and the maximum between 30° and 32°. At the optimum, the mycelium grows at a rate of approximately three-fourths of a centimeter per day; in other words, it requires about forty days for the mycelium to grow through 1 foot of soil. The results offer little hope of maintaining in the greenhouse a temperature very unfavorable to the growth of the fungus.

TABLE III. — *Effect of Temperature Variation on Rate of Mycelial Growth in Soil.*

TEMPERATURE, CENTIGRADE (DEGREES).	Number of Measurements.	Daily Growth in Centimeters.
5,	10	0
8.5,	10	0
14,	20	.26
18,	150	.37
21–22,	170	.50
23–25,	170	.61
25,	130	.63
25–26,	90	.68
25.5–26.5,	30	.69
26–27,	25	.74
30,	40	.25
32–33,	30	0
37.5,	10	0

Effect of freezing the Mycelium.

It is very important to know whether soil can safely be used in the benches after being frozen out of doors. The following tests were made to determine this point: —

Method. — Eight bottles, each containing 33 cubic inches of soil, were plugged, sterilized and inoculated with Cylindrocladium. After seven months the soil was thoroughly infested with the fungus, and probably contained all modifications of the mycelium which ever occur in the soil. Transfers were made and the fungus in all found to be alive. Then, before the ground froze in November, four of the bottles were exposed outside, one on top of the ground, one just under the surface, one 6 inches down, and one a foot below the surface. The other four were kept in the laboratory for controls. Some of these bottles were brought in each month of the winter to see whether the fungus was still alive.

The last test was made May 10, after the bottles had experienced the coldest winter on record in Massachusetts. The fungus was still living in the soil. Apparently, then, soil cannot be made safe by exposing it during the winter out of doors.

Thermal Death Point of Mycelium.

Anticipating soil sterilization by heat, the thermal death point for the mycelium was determined.

Method. — The same method was used as for determination of the thermal death point of spores, except that bits of agar containing mycelium were inserted into the sealed tubes, and after exposure for ten minutes to the desired temperature were transferred to sterile agar plates. If the mycelium was still alive it quickly began to spread to the agar. Temperatures between 42° and 55° C at intervals of 1° were tested.

Up to and including 48° the treatment seemed to have no effect on the mycelium. At 49° it was sometimes killed and sometimes not. It never grew after ten minutes' exposure to 50°. We may therefore consider 50° the thermal death point. It will be noticed that the thermal death points of mycelium and spores differ by only 1 degree. The mycelium tested contained, besides the ordinary white mycelium, also the dark bodies with thick walls which we have called chlamydospores and sclerotia. As was the case with spores, so also the mycelium may be killed by a longer exposure to a lower temperature. Based on an exposure during one hour, the thermal death point was found to be 48°.

DISSEMINATION.

In deciding on a method of controlling a disease it is of prime importance to find out how the pathogene is spread about, where it comes from, how it reaches the host. In the present case a threefold question is involved: (1) How did the fungus get into rose houses in the first place? (2) How is it spread from the houses of one rose grower to those of another? (3) On the premises of a single grower, how does it pass from house to house, bench to bench, or plant to plant? In the light of what has been learned concerning the life history and habits of the pathogene, we may undertake to answer these three questions.

1. Original Source of the Pathogene.

The fungus, from all that is known of its past history, is a native of America. Since it has been reported but a few times, it probably is not very common out of doors. As greenhouse roses are grown in the section of the country where it has been reported, it would not be far-fetched to imagine the fungus being carried into rose houses with rotted leaves, where it was able to adapt itself to parasitic life on the rose. It is not necessary to assume, then, that this is an imported pathogene. Early in the course of the investigation it was suspected that it might have been brought over from Europe on Manetti stocks, which are used almost exclusively by rose growers for grafting. The Manetti is moderately susceptible to the disease, as may be readily determined by examination of Manetti shoots coming from below the graft in a badly diseased house. Pure cultures have frequently been made from these shoots. Massey (1917) also made infection experiments and found Manetti roses susceptible. In the course of these investigations hundreds of Manetti stocks from Scotland were examined for lesions, numerous tissue plants were made, hundreds more were kept in moist chambers to bring out the fungus, and thousands of them watched carefully for a year after being planted in sterilized soil in order to see whether the disease developed. All results were negative, and up to the present we have no reason to suspect that the fungus is being imported on Manetti stock. It would be very helpful if we knew how widely the fungus is distributed over this country in its natural state, and whether it is being carried into the houses again and again. Various investigators have worked on the fungous flora of the soil and published lists of species isolated, but none of them mentions Cylindrocladium. This may indicate that it is only local in its distribution, or may be due merely to difficulties of isolating it. There seems to be little doubt that it infests the soil about rose houses where the disease occurs and where infested soil has been dumped out.

2. Spread from One Grower to Another.

Plants are continually being sent from one grower to another. Small cankers on these would be overlooked even if the sender was familiar with the disease. Not only could the mycelium be sent in the plant itself, but particles of soil adhering to the plants could easily carry it. It has been proved by laboratory tests that infested particles of soil may be kept dry for at least three months, and probably longer, without killing the mycelium. The disease may be spread in other ways, but this one would be sufficient to account for the present known distribution.

3. Local Dissemination.

There are a number of ways in which the fungus spreads from one part of a house to another, or from one plant to another. (a) It may grow for long distances through the soil and enter the plant below the surface of the soil. That infection can take place in this way has been repeatedly proved by setting clean plants in infested soil and thus producing the disease on them. (b) If the fungus is in the potting soil it would be effectually distributed in the beds when the plants were transplanted to them. (c) Where "own-root" plants are grown the soil in the cutting bench may be infested, and the disease is then carried with the cuttings when they are planted in the benches. (d) It is easily carried from one part of the house to another on tools, clothes and shoes of workmen.

(e) Insects, centipedes and worms carry the spores, as has been proved in the laboratory by permitting them to pass over sterile plates after being on dead twigs bearing spores. (f) The water used in watering the plants is usually driven from the nozzle with enough force to splash spores and bits of mycelium from the soil or débris on the ground up to the stems. Probably most of the stomatal infections above ground are started in this way.

The spores of many fungi are so light that they float around in the air and are wafted about by very light air currents. It does not seem likely that the spores of Cylindrocladium are carried about to any great extent in this way. They are bound together in solid heads of spores, which are probably too heavy for currents of air such as usually occur in rose houses. That they can be dislodged and blown some distance by *strong* air currents was proved in the laboratory by passing a strong current of air from a fan over spores growing on a dead rose stem, and exposing agar plates 1, 2 and 3 feet away. Colonies of the fungus developed on all of them, but it is hardly probable that so strong an air current would normally occur in rose houses. They could also be blown about on dust particles, but the soil in rose houses is rarely permitted to become dry enough to form dust.

OCCURRENCE OF TWO SPECIES OF CYLINDROCLADIUM ON ROSES.

During these investigations a second species of Cylindrocladium has frequently been isolated. It was first taken from the roots of a plant which had typical cankers on the crown. Later it was secured a number of times from crowns and from dead areas of the plant above the ground. It was commonly isolated directly from the soil in the rose beds, from the surface to 8 inches down. Except for its size, it resembles C. *scoparium* so closely that the writer was at first inclined to consider it but a dwarf variety of that species. The spores are only about one-third as large as those of C. *scoparium*. Although numerous isolations have been made, no transition forms between the two have been found. The small form has been grown through many generations in culture, and has remained constant on all media.

Infection experiments were carried out, but all attempts to produce the disease by the same inoculation methods as were used for the larger form gave only negative results. The fungus grows and produces spores on the dead tissue about wounds and on cut stubs, but seems to lack ability to spread to healthy tissue. The small form then appears to be a saprophyte, while the larger one is a parasite.

In order to determine whether there are cultural differences by which they could easily be distinguished, the two forms were grown simultaneously on five standard culture media. They show very marked diagnostic differences. Such differences in morphology, pathogenicity nd cultural characters are certainly marked enough to be considered specific rather than varietal. Since no species of Cylindrocladium other than *C. scoparium* has been described, a new name, *Cylindrocladium parvum*, is proposed for this small form.

The morphological differences and the cultural characters and differences of the two species are given in parallel columns below.

MORPHOLOGICAL CHARACTERS.

Since some morphological characters vary somewhat with the conditions under which they are grown, all measurements given below were taken from potato agar plates grown simultaneously under the same conditions, and each is the average of fifty measurements.

C. scoparium.	*C. parvum.*
Size of spores, 48.8 x 5.1 μ.	Size of spores, 16.8 x 2.5 μ.
Height of conidiophore, 291 μ.	Height of conidiophore, 130 μ.
Diameter of conidiophore stalk, 6.6 μ.	Diameter of conidiophore stalk, 4.25 μ.

CULTURAL CHARACTERS.

Most soil fungi can easily be grown on a great variety of artificial media. The characters of the colony differ markedly with the medium used, and very frequently species of fungi, like bacteria, can be distinguished more easily by macroscopic cultural characters than by microscopic morphological characters. Obviously, to grow each fungus on all the possible media, or even a great number of them, would be almost an endless task. Five common media, all easy of preparation, have therefore been adopted by the writer as standard for all diagnostic work. These five are (1) potato agar (acc. Thom. Bul. 82 U. S. D. A., Bureau of An. Industry); (2) sugar potato agar (the same as the potato agar except for addition of 3 per cent. of cane sugar); (3) gelatin (150 grams gold label to a liter of water); (4) sugar gelatin (same as above with addition of 3 per cent. of cane sugar); (5) Czapek's synthetic agar (acc. Waksman in Soil Sc. 2: 113). Petri dishes, each with a single colony started at the center, were used. They were kept in the diffused light of the laboratory at the ordinary laboratory temperature.

Every reference to a color in the description below refers to the color given under that name in Ridgway's "Color Standards and Nomenclature," 1912. Color "in reverse" in these descriptions refers to the color of the colony when examined from the bottom of the dish. This color may be due to (1) a pigment in the medium itself (extra-cellular), (2) intracellular pigments (*i.e.*, the natural color of the mycelium), or (3) very frequently it is due to a combination of the two. Sometimes a distinction is made between them, but for diagnostic work such a distinction usually adds difficulty instead of simplifying determination. Most emphasis is placed on those characters which appear within the first week after the colony is made. If one has to wait two or three weeks or longer for a character to appear, the long waiting makes diagnosis tedious, and one of the principal purposes of this method of diagnosis is defeated. The more important characters for distinguishing these two species are italicized. Many minor distinguishing characters are not mentioned.

POTATO AGAR.

C. scoparium.	*C. parvum.*
Growth only moderately good. Starts with abundant, perfectly white, raised, aerial mycelium, but soon falls flat at the center, which becomes covered with spores after two or three days. Always more or less aerial mycelium out toward the margin, which is rather coarse and tow-like. Not a decided color in reverse during the first week, but a dilute cream color to buff. At the end of the second week it turns to avellaneous or wood brown, and *after three weeks still darker, Rood's brown. Margin of colony crenulate or wavy.*	Only moderately good growth. Mycelium finer and denser than *C. scoparium,* perfectly white. Spores produced in great abundance. *The edge entirely throughout its growth remains very even and forms a perfectly round colony.* Practically *no color* — possibly a very faint buff — *develops in reverse even after three weeks' growth.*

SUGAR POTATO AGAR.

C. scoparium.	*C. parvum.*
Very rank growth, abundance of spores, entire plate covered in two weeks. Dense opaque color appears in reverse after three days; *vinaceous purple to hæmatite red at the edge, darkening to russet or chocolate at the center. At the end of a week a large central area appears almost black, but examined more closely shows various shades of reddish brown, chestnut and bay.* Entire reverse opaque after two weeks. The brown color is due to the extremely abundant production of sclerotia and chlamydospores on this agar.	Rank, white growth of a very much finer texture than *C. scoparium.* Abundant production of spores. *Color in reverse, white, or at most, only cream color at end of one week.* This is one of the best diagnostic characters. At the end of two weeks it has passed through gray and drab gray to a clear wood brown, with minute patches of army brown here and there which show chlamydospores under microscope. The red-brown colors of *C. scoparium* never appear.

GELATIN.

C. scoparium.	*C. parvum.*
Growth very poor, consisting of a thin covering of coarse radiating hyphæ. Very few spores. Stops growing after about ten days. Gelatin turned to a watery liquid which at the end of a week is *orange rufous, but gradually turns darker to Sanford's brown.* Liquefaction extends some distance beyond the margin of the colony.	Growth very scanty, so much so that it is necessary to look at the plate against a black background to see it at all during first week. Gelatin liquefied. No color at first, but becomes *dilute old gold by end of second week.* This medium is hardly suitable for distinguishing the two.

Sugar Gelatin.

C. scoparium.

Rank growth of coarse radiating aerial mycelium, but few spores. Gelatin liquefied. After about four days a *striking brilliant carmine color begins to appear in reverse, due to a pigment in the gelatin.* This gradually spreads to the whole plate *and becomes darker, an ox-blood red.* This is probably the best diagnostic cultural character for this species. The mycelium covers the plate in ten days.

C. parvum.

Fine tangled aerial mycelium and more abundant spore production than for *C. scoparium.* Gelatin liquefied. Covers entire plate in two weeks. *At the end of a week the colonies vary from Mars yellow to raw sienna in reverse, and at the end of two weeks have darkened to amber brown and Mars yellow.* The color during the entire development of the colony is in strong contrast to the carmine and ox-blood of *C. scoparium.*

Czapek's Agar.

C. scoparium.

Growth moderately good, aerial mycelium thin. Spores abundant. *At the end of a week the colors in reverse are much the same as for potato agar, — claret brown, russet or amber, with a brick-red color suffused through it. At the end of two weeks the center is practically black, fading through brown and red tints toward the margin.* The red color is due to a pigment in the medium; the brown, to the chlamydospores and sclerotia. Irregular edge.

C. parvum.

Finer and denser aerial growth of mycelium. *During the first week the reverse remains pearly white; later it changes to dilute wood brown, then Rood's brown and at the end of two weeks approaches Natal brown.* None of the red tints of *C. scoparium* ever appear. Margin much more even than that of *C. scoparium.* Abundant production of spores *in distinct concentric zones.*

Latin Description of Cylindrocladium parvum.

Cylindrocladium parvum n. sp. *Album effusum; conidiophoris erectis, base simplicibus, apice ternate vel dichotomice ramosis, 130 x 4.25μ; conidiis cylindraciis, medio obscure 1-septatis, hyalinis, 16.8 x 2.5μ.*

Hab. in caulibus emortuis et radicibus rosarum et in humo, Massachusetts in Amer. bor. — Simile C. scopario.

CONTROL.

Every method used in the control of any fungous disease is an application of one of four principles: (1) exclusion of the fungus, (2) eradication of the fungus, (3) protection of the host, or (4) immunization of the host. Although practically all the work of the present investigation has been on the second of these principles, there are possibilities of using all four of them in the control of rose canker. These four are first considered separately below in the order named, and finally a general scheme of treatment is recommended.

EXCLUSION OF THE PATHOGENE.

By exclusion we mean preventing a fungus from entering a given territory in the first place, whether this territory be a country, a State, a region or only one rose house. Since this disease seems to be pretty generally distributed over the country already it is obviously impossible to exclude it from the United States, and probably from any particular State or section. But it is entirely possible to exclude it from the house of a rose grower who finds that none of his plants are already affected, or where new houses are being erected at some distance from old ones. The whole practice, then, consists of taking every possible precaution against carrying any diseased stocks, cuttings or infested soil into the house. Every plant brought in should be carefully examined, and, if there are any suspicious cankers in the bark, it should be discarded. All new plants and cuttings should be taken whenever possible only from houses known to be free from the disease.

ERADICATION OF THE PATHOGENE.

By eradication we mean the absolute destruction or removal of the fungus from the rose beds or from the whole house, so that it is no longer present in the plants or in the soil, pots, débris, manure or anywhere else from which it can return to the plants. The practice of this method is of course necessary only when it has been impossible to exclude the pathogene and it has become established in the house. Up to the present this has proved to be the most successful principle applied to controlling canker.

The ultimate aim is to eradicate the fungus from the plant itself, but the application of direct methods, such as excision of cankers, pruning off of dead parts, or even absolute destruction of entire plants when cankers are found on them, is altogether useless because the soil all about the plants is infested. From the soil the fungus can grow back into the roses as fast as it can be cut out. Spraying or dusting is of course useless, also, because no fungicide can reach the mycelium in the inner tissues of the plant; and also it is not possible to cover the parts of the plant below the surface of the ground where infection commonly occurs. Obviously, then, eradication resolves itself into destruction of the pathogene in the soil; in other words, soil disinfection. Of the various methods of disinfecting soil only two have appeared to be at all practicable: (1) by heat, and (2) application of chemicals. Freezing, as previously mentioned, is not effective. Desiccation would take entirely too long. Other methods are either too expensive or too difficult of application. In the course of the present investigation both heat and chemicals have been successfully used.

Disinfection by Chemicals. Laboratory Tests.

Some of the chemicals which have been used in the past for disinfecting soil for the control of other fungous diseases are formaldehyde, sulfuric acid, copper sulfate, sulfur, lime-sulfur. The results obtained by the use of these same chemicals for other fungi could not be used directly in the present investigation because every fungus differs in its resistance to a given chemical. It was first necessary to determine what concentration and what quantity of solution per cubic foot was needed to kill the fungus. These facts could be determined more accurately and conveniently in the laboratory than in the greenhouse. The method used in all these tests was as follows: —

Method. — Milk bottles, each containing 33 cubic inches of soil, were steam sterilized and inoculated from pure cultures of the fungus. When the soil was entirely infested (requiring from twelve days to three weeks) it was stirred into a loose condition with a sterile glass rod, and the proper amount of chemical in solution, at the strength to be tested, poured in under aseptic conditions. Since the soil did not dry out as rapidly in these bottles as it would under natural conditions in the greenhouse, it was emptied into sterilized porous flowerpots after a few hours. It was found after several trials that the pots dried out too rapidly if left in the open laboratory. Thereafter they were covered with bell jars which were tilted enough to allow free circulation of air beneath them, and the length of the drying process could then be regulated. After eight to ten days in the pots, clods of the soil were transferred from various portions of the pots to sterile agar plates. If the fungus was still alive it spread to the agar; otherwise there was no growth whatever from the clods. At first, the solutions were applied at the rate of 1 gallon to the cubic foot of earth. Afterwards, 2 gallons per cubic foot were used. When dry chemicals, such as sulfur, were tested the required amount was thoroughly stirred into the infested soil of the bottles with a sterile rod and no water added.

Formaldehyde. — First tests were at the rate of 1 gallon per cubic foot at the following concentrations: 1–500 (1 part of commercial formaldehyde to 500 parts of water), 1–400, 1–300, 1–200 and 1–100. None of these concentrations gave complete success. On the transfers from the last two, however, only a few of the clods contained living mycelium. This indicated a lack of complete penetration by the solution. In the next series of tests the same concentrations at the rate of 2 gallons per cubic foot were used. The 1–100 and 1–200 then gave absolute control, while the 1–300 usually did; but occasionally a single clod developed a mycelium on the agar. The death point concentration lies somewhere between 1–200 and 1–300. But to be well within the margin of safety, 1–200 (1 pint of commercial formaldehyde solution to 25 gallons of water) was decided upon as the best strength to use in the greenhouse.

Sulfuric Acid. — This chemical has been successfully used in the past in the control, particularly, of certain root diseases of nursery trees. At the rate of 2 gallons per cubic foot, concentrations of 1, 2, 3, 4, 5 and 8 per cent. were used. The 5 per cent. solution killed most of the mycelium, but not all of it. The 8 per cent. killed all of it. The death point concentration lies between 5 and 8 per cent., but such a high concentration is hardly practicable in the rose house, and the exact point was not determined.

Copper Sulfate. — Concentrations of 1, 2, 3, 4, 5 and 10 per cent. were used at the rate of 2 gallons per cubic foot. The 5 per cent. seemed hardly to check the fungus, but 10 per cent. proved entirely effective. Such a high concentration seemed prohibitive for application to soil, and no more accurate determination was made.

Lime-sulfur. — This mixture proved to be worthless, even when applied at a concentration of 1 part of commercial product (32° Baume) to 10 gallons of water, and at the rate of 2 gallons per cubic foot.

Dry Sulfur. — Finely ground sulfur flour was added to the soil and thoroughly stirred in. First, 10 grams per bottle were used, and when that proved to be ineffective 10 grams more were added, etc. All results were negative, even up to the rate of 7 pounds of sulfur to a cubic foot of soil. This test was performed at a laboratory temperature of 19° to 24° C. Perhaps if higher temperatures had been used the sulfur would have been more effective. Dry sulfur seems to be worthless at the temperatures tested.

Soot. — There is an idea prevalent among florists that soot has fungicidal value, but plant pathologists seem never to have made any extensive experiments with it. The same method and rates as for dry sulfur were tried. At the rate of 4 pounds per cubic foot soot did not kill the fungus, but at the rate of 7 pounds no growth of the pathogene occurred.

Of all the chemicals tried, formaldehyde seemed to be the only one which would give control at concentrations which could safely be used on the soil.

Greenhouse Tests with Formaldehyde.

The greenhouse tests on the use of formaldehyde were begun before the laboratory tests were completed, and at a time when it appeared that a concentration weaker than 1 pint to 25 gallons would be sufficient. As a result, the tests on a large scale were made with a concentration of about 1 pint to 40 gallons, but, on the other hand, more solution was applied per unit of soil. Two houses, each capable of growing more than 1,000 rose plants, were thoroughly soaked with the solution. One of the houses contained raised benches; the other, ground beds. Both had previously grown diseased roses. The soil was replaced by soil from outside the houses before sterilization. In the light of what we now know of the habits of Cylindrocladium, it is safe to assume that this soil was infested, because soil from the benches in previous years had been thrown out near it. After soaking the soil thoroughly the houses were closed. Fumes of formaldehyde were so strong in the closed houses that it was not possible to remain in them. After the soil had dried sufficiently both houses were planted with roses which had been potted in soil sterilized with steam, and which had been kept under conditions as sterile as possible. Three months after planting, no disease had appeared in either house. Soon afterward it began to appear in the house with the ground beds, and gradually increased until, almost a year after planting, it was generally prevalent throughout the house. In the bench house, however, no disease has as yet been found, although plant-to-plant inspections have been made frequently throughout the year. The fact that a concentration of formaldehyde weaker than 1 pint to 25 gallons controlled the disease in the bench house is probably due to the longer action of the more concentrated fumes, and probably, also, partly to the greater amount of the solution applied. The lack of control in the ground bed house can be easily explained in the light of our studies on the depth of penetration of the mycelium in the soil. The surface soil was disinfected, but it was not possible to disinfect it down as far as the mycelium grows. After the formaldehyde had evaporated the deep mycelium began to grow upward, and during that period the plants remained healthy; but, after the mycelium had grown up to the surface again, the cankers began to appear and the roses became as badly affected as before the house was treated. Two conclusions may be drawn from this experiment: (1) the soil can be disinfected effectively by the use of formaldehyde, and (2) ground beds cannot be sterilized by this method.

Disinfection by Heat. Laboratory Tests.

The feasibility of destroying any fungus by application of heat to the soil manifestly depends, first of all, on the thermal death point of all stages of that fungus. As has previously been described, this point for Cylindrocladium was found to be 50° C. This comparatively low death point indicated that the soil could be readily disinfected by steaming, because a temperature much higher than 50° C. can be easily obtained by the use of steam.

Time required to disinfect Soil by steaming. — This was further confirmed by the following tests: —

Method. — Sterile Petri dishes were filled with soil which was thoroughly infested with mycelium. After removing the lids they were subjected to steam at a temperature of 90° to 95° in an Arnold sterilizer for the desired length of time. The lids were then replaced and the soil allowed to cool, when clods of it were transferred to agar plates as described above. Exposures of five, ten, fifteen, twenty and thirty minutes were tried.

No mycelium appeared on any of the transfers, even after five minutes' exposure. Shorter periods of exposure were not tried because of the uncertainty of securing penetration by steam in less than five minutes. But, to determine what effect shorter exposures would have on mycelium, tests were made by the sealed tube method described for thermal death point tests. In these tests the mycelium was killed in less than one minute when exposed to a temperature of 95° C.

From these tests we may conclude that soil can be disinfected by steam in less than a minute if penetration is obtained. Apparently effectiveness is limited only by the time required for the steam to penetrate every particle of the soil.

Greenhouse Tests of Disinfection by Heat.

Heat may be applied to the soil by steam or by hot water. The first method has been in use in the greenhouses for the disinfection of the soil used in potting since the beginning of this investigation. Perforated steam pipes were laid a foot apart in a large pit. Soil a foot deep or more was piled over them and the steam turned into the pipes. Burlap or other coverings may be used to cover the soil and make it retain more of the steam. Soil thermometers were used to determine the temperature. It is only necessary to keep the temperature above 50° C. for ten minutes. A higher temperature, of course, makes for additional safety. The one or two hours of heating frequently recommended for other diseases is only wasted time and expense, being entirely unnecessary for this fungus. Thousands of plants have been potted in soil disinfected in this way during the last year, and canker has never appeared on any of them. No doubt other methods of steam disinfection, such as the inverted pan method, would be equally effective. Either method could probably be used just as effectively on the benches, but the formaldehyde treatment is efficient, and quicker and easier of application.

If there is any reason to suspect the presence of the fungus in the manure which is used to mulch the beds it may be disinfected in the same way as the potting soil. Soil for the cutting bench may also be treated in the same way.

The second method of applying heat — by the use of boiling water — is now being tested. It should be just as effective as steam, and at the same time much more rapid. The boiling water is forced through the water pipes ordinarily used in the house, and is applied to the soil through a hose with a long nozzle and a handle which will not become heated. The water should be applied until a thermometer inserted into the soil at any point and at any depth registers above 50° C. Higher temperatures make for additional safety. This method has the disadvantage of leaving the soil in poorer condition for working. The hot-water method is still in the experimental stage, and is not far enough along to warrant any recommendations.

Disinfection of Pots, Tools, etc.

In starting new houses with clean plants and clean soil, it is very essential that everything which is used should be free from any form of inoculum. The first danger is from pots which have been previously used, and which are apt to contain mycelium or spores in the particles of earth which still cling to them. They can be sterilized by immersing in boiling water for ten minutes. Steaming is just as effective. The method used is simply a matter of convenience.

Usually a grower, when he finds disease in his houses, finds it impracticable to destroy all his roses and start all over again. Therefore he retains some of his old houses and starts disinfection operations on one or more, from which he has removed all the plants. This inevitably results in the constant danger of carrying some infested soil or parts of plants from the infested to the clean houses. Every possible precaution should be taken to guard against this, because a failure here means that the work must all be done again. All sorts of tools offer an easy means of conveying the inoculum. Whenever possible an entirely different set of tools should be used in the clean houses, and no tools from the other houses brought in under any conditions. But, if this is not possible, the next best alternative is to sterilize the tools before bringing them in. The method of sterilizing them is not so important as thoroughness. They may be dipped in boiling water, steamed, or a barrel of Bordeaux mixture or formaldehyde — preferably stronger than 1 pint to 25 gallons in this case — may be used for soaking the tools.

It may be necessary to sterilize other things besides pots and tools, e.g., boots and clothes of workmen. Every grower, after learning the habits of the pathogene, must decide for himself on the best way, under his own conditions, of keeping his houses clean.

PROTECTION OF THE HOST.

By protection we mean the placing of a barrier between a plant and a pathogene which would otherwise attack it and cause disease. This is well exemplified in the extensively used practice of spraying plants, the fungicide forming a poison barrier through which the fungus cannot penetrate. The humicolous habit and underground method of attack of the canker fungus seem to preclude any hope of important benefit from spraying. There is one place in the propagation of roses, however, where a fungicidal covering might be beneficial. Scions and cuttings should, whenever possible, be taken from houses known to be clean. If they are taken from houses in which the disease occurs there is always a possibility of spores being lodged on them, even where lesions have not as yet appeared. To either wash off and kill these spores or, at least, to prevent germination where they are, it has been the practice during this investigation to dip all such cuttings in a fungicide before grafting or planting.

Comparative Value of Different Fungicidal Coverings.

In order to find the best fungicide to use for dipping, and also to secure data for use in case spraying should be found advisable at any time, the comparative value of a number of fungicides was tested in the laboratory.

Method. — Glass slides were sprayed with the fungicide to be tested and permitted to dry for varying periods of time. Then spores of the fungus in a drop of water were transferred to the center of the sprayed slide, which was then kept in a moist chamber for twenty-four hours. Checks on unsprayed slides were always made at the same time. Percentages of germination were counted at the end of twenty-four hours, and observations were taken for several days to see if there was any further development; but none of the results in these tests were modified by later observations. When a dry fungicide was used it was dusted on to the slide without water. All checks in these tests germinated over 95 per cent.

Lime-sulfur. — Concentrations of 1–10, 1–30 and 1–50 commercial lime-sulfur solution were used. The 1–50 concentration proved to be useless from the start. The 1–30 seemed to check germination at first, but after it had been on the slide four or five days over 50 per cent. of the spores germinated. The 1–10 concentration entirely prevented germination when fresh, but after a week the control was erratic, with over 50 per cent. germination on some of the slides. Commercial lime-sulfur seems to be useless for control of this fungus.

Dry Sulfur Flour. — Slides were very heavily dusted and the germination tests made at about 25° C. The presence of the sulfur had no effect whatever on the spores. They germinated just as well as the checks. Dry sulfur appears to be even less effective than the lime-sulfur.

Ammoniacal Copper Carbonate. — This fungicide prevented germination twenty-four hours after being dried, but when tried a week later was only 25 per cent. efficient. This would hardly be a safe fungicide.

Lime. — Milk of lime sprayed on the slides from an atomizer prevented germination from the first, and was just as effective as Bordeaux. Milk of lime is not suitable for dipping cuttings. The lime test was made with a different end in view.

Bordeaux Mixture. — This fungicide was made up at a strength of 4–4–50. Germination tests were made every day for twenty-one days after the slides were sprayed. No germination occurred in any of these tests. These fungicidal tests clearly indicate Bordeaux mixture as the most suitable solution for dipping cuttings.

Treatment of the Walks in the House.

Undoubtedly the walks between the benches of a house which has previously grown diseased roses are infested with the pathogene. One could easily think of a great many ways in which small particles of soil from the walks could be carried into the benches. It is therefore necessary either to keep the fungus killed out of the surface of the walks by repeated applications of some fungicide or to cover the walks with some substance which will be a barrier through which it cannot pass up to the benches. In the beginning of this investigation the walks were kept sterile by frequent applications of formaldehyde. This proved unsatisfactory because the fumes of formaldehyde often injure the roses, producing dead spots on the leaves. This was abandoned and a search begun for something more suitable. Up to the present, lime gives the best promise of making a satisfactory barrier. Sterile bottle tests show that the mycelium will not grow in soil containing air-slaked lime at the rate of 1¼ pounds per cubic foot. Neither will spores germinate in the presence of lime. Until something more satisfactory is found it is recommended that all walks in the houses be kept covered with lime. Not only will this furnish an effective barrier to the fungus coming up from below, but it will also prevent growth of spores and other inocula brought in from other houses on the shoes of workmen and visitors.

Immunization of the Host.

By immunization we mean either the development of varieties of roses which are immune, — at least highly resistant, — or rendering them immune by injection or feeding through the roots with some chemical. No work has been done along either of these lines in regard to rose canker. From the first it has been noticed that some varieties of roses are more susceptible than others. No doubt in the course of time desirable varieties will be found or developed which will not suffer from canker. How soon that will be no one can predict. A rose breeder of wide national reputation told the writer that he had spent most of his life producing four or five varieties of roses. It is a long process, and until such varieties are developed it will be necessary to resort to such emergency measures as have been described in this bulletin.

Summary of Control Measures.

In the light of all that we know about rose canker and its causal pathogene the following measures are recommended for its control: —

1. Carefully inspect the rose house to see if canker is present. If not, employ every means to prevent its entering, — import as few roses as possible from other houses; examine carefully every plant brought in; reject any with suspicious dead areas in the bark.

2. If it is present on the roses it cannot be eradicated from the infected plants. The only hope lies in starting new plants from clean cuttings in clean soil, and guarding against infection at every step in the plant's development.

3. Dip the cuttings in Bordeaux mixture.

4. Sterilize the pots by dipping for ten minutes in boiling water.

5. Sterilize the potting soil and cutting bench soil by steaming to a temperature of over 50° C. for ten minutes or more. Suspected manure should be treated in the same way.

6. Use raised benches, not ground beds.

7. Remove old soil if diseased roses have been grown in it, and soak the benches thoroughly with (1) formaldehyde at the rate of 1 pint to 25 gallons, or (2) boiling water.

8. Sterilize the bench soil by one of these two methods. If formaldehyde is used, apply at the rate of 2 gallons per cubic foot. If boiling water is used, apply until every part of the soil is heated above 50° C.

9. Use a different set of tools in the clean house, or sterilize all tools before bringing them in.

10. Keep the walks in all houses covered with lime.

LITERATURE CITED.

Morgan, A. P., 1892. "Two New Genera of Hyphomycetes." Bot. Gas. 17: 190–192.

Ellis, J. B., and Everhart, B. M., 1900. "New Species of Fungi from Various Localities, with Notes on Some Published Species." Bul. Tor. Bot. Club 27: 49–64.

Massey, L. M., 1917. "The Crown Canker Disease of the Rose." Phytopathology 7: 408–417.

Supporting Dyslexic Learners in the Secondary Curriculum
Moira Thomson, MBE

DYSLEXIA:
INFORMATION FOR GUIDANCE, PASTORAL &
BEHAVIOUR SUPPORT TEACHERS

First published in Great Britain by Dyslexia Scotland in 2013

Second edition for schools in England published in 2017 by CPD Bytes Ltd

ISBN 978-1-912146-32-1

This booklet is 1.8 in the series
Supporting Dyslexic Learners in the Secondary Curriculum (England)

Supporting Dyslexic Learners in the Secondary Curriculum   Moira Thomson, MBE

Complete set comprises 25 booklets

1.0   Dyslexia: Secondary Teachers' Guides
1.1   Identification and Assessment of Dyslexia at Secondary School
1.2   Dyslexia: Underpinning Skills for the Secondary Curriculum
1.3   Dyslexia: Reasonable Adjustments to Classroom Management
1.4   Dyslexia: Role of the Secondary School SENCo (Dyslexia Specialist Teacher)
1.5   Partnerships with Parents of Secondary School Students with Dyslexia
1.6   Dyslexia: ICT Support in the Secondary Curriculum
1.7   Dyslexia and Examinations (Reasonable Adjustments & Access Arrangements)
1.8   Dyslexia: Information for Guidance, Pastoral & Behaviour Support Teachers
1.9   Dyslexia: Study skills and Interventions for the secondary curriculum
1.10   Dyslexia: Role of the Teaching Assistant
1.11   Dyslexia: Co-occurring & Overlapping Issues (Specific Learning Difficulties) NEW

2.0   Subject Teachers' Guides
2.1   Dyslexia: Art & Design Subjects
2.2   Dyslexia: Drama (Performing Arts; Theatre Studies)
2.3   Dyslexia: English (Communication)
2.4   Dyslexia: Home Economics (Child Development; Food & Nutrition)
2.5   Dyslexia: ICT Subjects (Business Subjects; Computer Science)
2.6   Dyslexia: Mathematics (Statistics)
2.7   Dyslexia: Modern Foreign Languages
2.8   Dyslexia: Music
2.9   Dyslexia: Physical Education (Sports; Games; Dance)
2.10   Dyslexia: Science Subjects (Biology; Chemistry; General Science; Physics)
2.11   Dyslexia: Social Subjects (Economics; Geography; History; Citizenship Studies; Philosophy; Religious Studies)
2.12   Dyslexia: The Classics (Classical Greek; Latin; Classical Civilisations) (2013)
2.13   Dyslexia: Media Studies NEW
2.14   Dyslexia: Social Sciences (Anthropology, Archaeology; Humanities; Psychology; Sociology) NEW

Foreword by Dr. Gavin Reid, formerly senior lecturer in the Department of Educational Studies, Moray House School of Education, University of Edinburgh. An experienced teacher, educational psychologist, university lecturer, researcher and author, he has made over 1000 conference and seminar presentations in more than 40 countries and has authored, co-authored and edited many books for teachers and parents.

ACKNOWLEDGEMENTS

Moira Thomson would like to thank the following for making possible the original publication of this important series of booklets:

- ✦ Dyslexia Scotland for supporting the publication and distribution of the original editions of these booklets

- ✦ The Royal Bank of Scotland for an education grant that funded Dyslexia Scotland's support

- ✦ Dr Gavin Reid for his encouragement over the years – and for writing the Foreword to these booklets

- ✦ Dr Jennie Guise of DysGuise Ltd for her support and professional advice

- ✦ The committee of Dyslexia Scotland South East for their support

- ✦ Alasdair Andrew for all his hard work and unfailing confidence

- ✦ Colleagues Maggie MacLardie and Janet Hodgson for helpful comments

- ✦ Cameron Halfpenny for proof reading and editing these booklets

- ✦ Current and former students, whose achievements make it all worthwhile

Moira Thomson MBE
2017

FOREWORD by Dr Gavin Reid

The Dyslexia booklets written by Moira Thomson have been widely circulated and highly appreciated by teachers throughout Scotland and beyond. I know they have also been used by teachers in a number of countries and this is testimony to the skills of Moira in putting together these booklets in the different subject areas of the secondary school curriculum.

It is therefore an additional privilege for me to be approached again by Moira to update this Foreword to the compendium of books developed by Moira in association with Dyslexia Scotland.

These updated guides are for all teachers - they contain information that will be directly relevant and directly impact on the practice of every teacher in every secondary school in the country. It is heartening to note that the guides again provide a very positive message to readers. The term dyslexia is not exclusive to the challenges experienced by learners with dyslexia, but there is now a major thrust towards focussing on the strengths and particularly what they **can** do - and not what they 'can't do'. It is important to obtain a learning profile which can be shared with the student.

Moira encapsulates these points in these updated booklets. The focus is on supporting learners and helping them overcome the barriers to learning. At the same time it is important that learners with dyslexia, particularly in the secondary school develop responsibility for their own learning. The acquisition of self-sufficiency in learning and self-knowledge is an important aspect of acquiring efficient learning skills for students with dyslexia. It is this that will stand them in good stead as they approach important examinations and the transition to tertiary education and the workplace. For that reason these guides are extremely important and need to be available to all teachers. Moira ought to be congratulated in endeavouring to achieve this.

The breadth of coverage in these guides is colossal. Moira Thomson has met this immense task with professionalism and clarity of expression and the comprehensiveness of the guides in covering the breadth of the curriculum is commendable.

As well as including all secondary school subjects the guides also provide information on the crucial aspects of supporting students preparing for examinations, the use of information and communication technology, information for parents, details of the assessment process and the skills that underpin learning. It is important to consider the view that learners with dyslexia are first and foremost 'learners' and therefore it is important that their learning skills are developed fully. It is too easy to place the emphasis on developing literacy skills at the expense of other important aspects of learning. The guides will reinforce this crucial point that the learning skills of all students with dyslexia can be developed to a high level.

The guides do more than provide information on dyslexia; they are a staff development resource and one that can enlighten and educate all teachers in secondary schools. I feel certain they will continue to be warmly appreciated. The guides have already been widely appreciated by teachers and school management as well as parents and other professionals but the real winners have been and will continue to be the **students** with dyslexia. It is they who will ultimately benefit and the guides will help them fulfill their potential and make learning a positive and successful school experience.

Dr Gavin Reid, April 2016

WHAT IS DYSLEXIA?

Dyslexia is widely recognised as a specific difficulty in learning to read.

Research shows that dyslexia may affect more than the ability to read, write and spell – and there is a growing body of research on these 'co-occurring' factors.

The Rose Report[1] identifies dyslexia as *'a developmental difficulty of language learning and cognition that primarily affects the skills involved in accurate and fluent word reading and spelling, characterised by difficulties in phonological awareness, verbal memory and verbal processing speed.'*

Dyslexia is a learning difficulty that primarily affects the skills involved in accurate and fluent word reading and spelling.

Characteristic features of dyslexia are difficulties in phonological awareness, verbal memory and verbal processing speed.

Dyslexia occurs across the range of intellectual abilities.

It is best thought of as a continuum, not a distinct category, and there are no clear cut-off points.

Co-occurring difficulties may be seen in aspects of language, motor co-ordination, mental calculation, concentration and personal organisation, but these are not, <u>by themselves,</u> markers of dyslexia.

A good indication of the severity and persistence of dyslexic difficulties can be gained by examining how the individual responds or has responded to well-founded intervention.

Rose Report page 10

Dyslexia *exists in all cultures and across the range of abilities and socio-economic backgrounds. It is a hereditary, life-long, neuro-developmental condition. Unidentified, dyslexia is likely to result in low self-esteem, high stress, atypical behaviour, and low achievement.*[2]

Estimates of the prevalence of dyslexia vary according to the definition adopted but research suggests that dyslexia may significantly affect the literacy attainment of between 4% and 10% of children.

---

[1] Rose, J (2009)Identifying and Teaching Children and Young People with Dyslexia and Literacy Difficulties DCFS Publications - independent report to the Secretary of State for Children, Schools &Families June 2009. http://webarchive.nationalarchives.gov.uk/20130401151715/http://www.education.gov.uk/publications/eOrderingDownload/00659-2009DOM-EN.pdf

[2] From Scottish Government working definition of dyslexia http://www.gov.scot/Topics/Education/Schools/welfare/ASL/dyslexia[3] SEND Code of Practice 0-25 https://www.gov.uk/government/uploads/system/uploads/attachment_data/file/398815/SEND_Code_of_Practice_January_2015.pdf

TEACHERS' RESPONSIBILITIES RE LEARNERS WITH DYSLEXIA

References: Part 6 of the Equality Act 2010; Part 3 of the Children and Families Act 2014

All children/ young people are entitled to an appropriate education, one that is appropriate to their needs, promotes high standards and the fulfillment of potential - to enable them to:

- achieve their best
- become confident individuals living fulfilling lives, and
- make a successful transition into adulthood, whether into employment, further or higher education or training

SEND Code of Practice 0-25[3]

All schools have duties towards individual children and young people to identify and address their Special Educational Needs/Disability (SEND). Dyslexia that has a substantial, long-term, adverse impact on day-to-day learning may be both SEN and a disability.

Teachers' responsibilities for meeting the needs of dyslexic learners are the same as those for all students, and should include approaches that avoid unnecessary dependence on written text.

Teachers have a responsibility to provide a suitably differentiated subject curriculum, accessible to all learners, that provides each with the opportunity to develop and apply individual strengths – and to ensure that learners with SEND get the support they need to access this. Rose[4] suggests that all teachers should have 'core knowledge' of dyslexia characteristics – to help them to make adjustments to their practice that will prevent discrimination and substantial disadvantage.

Dyslexia may be difficult for some teachers to identify in a subject context – some think that dyslexia has little or no impact in their subject – others believe that dyslexia will have been resolved at primary school. The impact of unsupported dyslexia on learning in secondary subject classrooms may be profound, and result in a mismatch between a student's apparent subject ability and the quality (and quantity) of written work.

While subject teachers are not expected to diagnose dyslexia without specialist input - they should be aware of its core characteristics and likely manifestations in the classroom so they may refer students for assessment. Many schools have checklists and questionnaires in place to help teachers identify possible SEN and subject teachers should use these and follow school procedures when they suspect that dyslexia might be present.[5]

---

[3] SEND Code of Practice 0-25
https://www.gov.uk/government/uploads/system/uploads/attachment_data/file/398815/SEND_Code_of_Practice_January_2015.pdf
[4] Rose Report (2009) page 17
[5] A version of a Dyslexia Indicators Checklist for secondary age students is provided at the end of this booklet[6]
Use a Dyslexia Checklist when students exhibit behavioural difficulties to help identify dyslexic issues.

IMPACT OF DYSLEXIA ON LEARNING

Dyslexia may result in difficulties with processing language-based information - either auditory or visual processing difficulties, or both.

These processing difficulties often manifest as behavioural issues, especially in teenagers.

At least one person in ten is thought to be dyslexic to some degree and the learning of 4% could be severely affected by their dyslexia. Some people think that there are more males than females with dyslexia, but it is probable that many girls are not identified as dyslexic at school because they can compensate better than boys of the same age due to differences in the way they process language. It is also claimed that young girls exhibit less attention-seeking behaviour than boys so any dyslexia is less likely to be identified in school.

It is important that secondary teachers consider dyslexia in the context of the subject curriculum. In all subject classes there will be a need for teachers to make provision to meet a wide variety of strengths and additional support needs, not all of which will be linked to dyslexia. However, teaching and learning strategies that are appropriate for students with dyslexia can be effective for all learners.

DYSLEXIA - TRANSITION PRIMARY- SECONDARY
Reference: NASEN Transition Guide (2014) http://www.sendgateway.org.uk/resources.transition.html

The SENCo will be closely involved in transition arrangements from primary to secondary school in order to anticipate circumstances that might trigger learning, social, emotional or behavioural issues. Special or enhanced transition arrangements are often put in place for children identified (by primary staff and other professionals) as those who might find changing schools particularly challenging. It is common practice for children identified as dyslexic to be offered enhanced transition support by the secondary specialist dyslexia teacher, while the SENCo may arrange for a Student (or Behaviour) Support specialist to provide support to those who may present challenging or withdrawn behaviour when they move from a familiar primary school to a larger, busier secondary school.

The secondary SENCo often attends review meetings held in the later stages of primary and a specialist dyslexia teacher may work closely with primary staff to identify possible barriers to learning and explore the nature of any reasonable adjustments/interventions/individual support that will be needed for students with dyslexia to access the secondary curriculum at an appropriate level. This information is then used to inform subject teachers of any strengths and possible difficulties that dyslexic students may show in their subject curriculum and to suggest strategies that should be adopted to overcome barriers and ensure access to curricular materials and activities.

While many learning support teachers are concerned with learners' access to the subject curriculum - behaviour support teachers are more likely to consider the possible impact of transition on individual children emotionally and socially and look for ways of minimising any negative aspects of this. Dyslexic children may respond best to a combination of the support offered from both 'specialisms' at times of transition from one stage to the next.

SOME 'BEHAVIOURAL' CHARACTERISTICS OF DYSLEXIA

Learners with dyslexia constantly meet barriers to learning across the curriculum and may become discouraged very quickly due to lack of initial success in subject classes. This can result in subject teachers assuming that students are inattentive or lazy, when they are actually working much harder than their classmates, but producing very little. For students with dyslexia the experience of success may be rare, if not totally absent. In addition to struggling with literacy, they may:

- lack self-confidence
- have a poor self-mage
- fear new situations
- confuse written and verbal instructions
- appear to avoid set work
- be very disorganised
- lack stamina

For example, student with dyslexia may fully understand the subject teacher's spoken introduction to a topic but be unable to follow the written instructions to complete class activities.

IDENTIFICATION OF DYSLEXIA AT SECONDARY

Many secondary teachers assume that any dyslexia will have been identified and assessed at primary school, and that relevant information will be passed to them as part of transition arrangements. But there are some aspects of dyslexia that do not become apparent until students begin to experience difficulties within the secondary curriculum -perhaps having reached a stage where they are no longer able to use their strengths to compensate for dyslexic difficulties. In some cases, this may not be until students are about to sit timed examinations.

Dyslexia may not be identified until after students transfer to secondary school for a number of reasons which result in them experiencing barriers to learning:

- differences between the primary classroom and the busy secondary school timetable, cause dyslexic problems to emerge
- the move to secondary school has eliminated many of the support strategies that students with 'hidden' dyslexia developed at primary school to mask that they were having problems
- some higher order dyslexic difficulties may not appear until the demands of the secondary curriculum cause a student's coping strategies to collapse
- the time aspect of the secondary timetable often creates problems for students

with dyslexia accustomed to having all the time they need to complete set tasks in the primary classroom

- a mismatch between a student's apparent ability and the quality (and quantity) of written work emerges in some subjects

Some learners find it difficult to cope in the classroom and their behaviour presents a great challenge to teachers. They may be identified as having social, emotional or behavioural problems. These learners may be dyslexic - but teachers - and other people around them - have not understood that behavioural issues may be manifestations of dyslexia.

It will not be possible for the SENCo or specialist teacher to predict if/when students will 'suddenly' reveal dyslexic difficulties through their behaviour- but it is very important that that subject teachers are aware that this is possible. **It is important for teachers to understand that strategies employed for dealing with behavioural issues may not be effective unless any underlying dyslexia is also addressed.[6]**

---

[6] Use a Dyslexia Checklist when students exhibit behavioural difficulties to help identify dyslexic issues.

CHARACTERISTICS OF DYSLEXIA

Each student with dyslexia will have a very distinctive profile of strengths, difficulties and learning style/preference, so comparisons with other known students with dyslexia **may** not be useful, although there is often some common ground. Many, but not all, of the following characteristics may be present - learners will have their own individual combination of strengths and difficulties.

Barriers to learning: students with dyslexia may
- underachieve academically
- perform well orally or in practical activities but find reading/writing difficult
- be considered clumsy and disorganised
- appear restless, with poor concentration span
- seem inattentive, forgetful, easily tired
- have a low tolerance of their own lack of achievement
- have low self-esteem

Teachers should be aware that
- dyslexic difficulties can range from mild to severe and individual profiles will show both strengths and, sometimes, unexpected weaknesses
- dyslexia can occur at any level of intellectual ability
- learners with dyslexia often have natural talents, creative abilities and vision
- learners with dyslexia often display differences and experience difficulties in education, some of them hidden

HIDDEN DYSLEXIA

Dyslexia is often hidden, masked by a student's high ability or by distracting behaviour - even deliberately concealed by teenagers who are desperate not to be 'different' from their peers. Some students' true levels of ability may be masked by dyslexia - when they perform at the expected level but are actually being limited by underpinning dyslexic difficulties, resulting in unidentified underachievement and, perhaps, a curriculum that lacks challenge and fails to stimulate them.

Many learners with dyslexia become discouraged by constantly meeting barriers to learning, perhaps resulting in an assumption that they are inattentive or lazy, when they are actually making much more effort than their classmates. Because their dyslexia has not been recognised - or has been discounted by teachers who assume that they have somehow 'grown out of it' - some students believe that they really are lazy or incapable of concentrating as required. The resulting low self-esteem can lead to mental health issues, sometimes requiring therapeutic input by educational psychologists and/or medical professionals.

DYSLEXIA AND UNDERACHIEVEMENT

The characteristics of underachievers with dyslexia often reflect the difficulties they encounter - e.g. those from disadvantaged backgrounds may have low self-esteem and poorly developed study skills, resulting in failure to persevere at tasks and hostility to school and formal learning. Some students with dyslexia may be very able orally and mature in conversation but unable to write at length, with poor spelling and handwriting, leading teachers to underestimate their effort, ability and interest level. Patterns of underachievement by learners with dyslexia often include:

- high cognitive ability but low self-esteem
- poor work habits and unfinished tasks
- an apparent inability to concentrate
- lack of effort in some subjects but often an intense interest/skill in another
- a skill deficit in one area or subject
- a negative attitude towards self and age-peers
- manifestations of emotional frustration
- failure to respond to appropriate stimulation

Because of persistent failure in some aspects of the curriculum, underachieving learners with dyslexia may manifest either aggressive or withdrawn behaviour alongside some characteristics of high ability and dyslexia.

Features of the aggressive behaviour are:

- rejection of set tasks
- lack of co-operation
- disruption and alienation of others

Some students with dyslexia deliberately seek confrontation with their teachers in order to avoid set work or to be removed from the class.

Withdrawn behaviour includes:

- lack of communication, a preference for working alone
- daydreaming, little set work undertaken
- apparent lack of concern about attitude or behaviour

Underachievers with dyslexia can be hard to identify because they may be both experienced and skilled at hiding their dyslexia, which often results in their actual ability being unrecognised. They may not have been identified as having abilities or learning needs that are in any way different from those of most of their age-peers - and teachers may have neither time nor the inclination to search for hidden abilities in uncooperative learners.

Learning characteristics of underachievers with dyslexia: they may

- disguise their level of ability to conceal dyslexia and gain peer acceptance
- prefer to be accused of lack of effort and concentration to admitting difficulties
- reject set tasks and fail to respond to teachers' instructions

- fear public failure, inhibiting attempts in new areas
- be frustrated with inactivity, and lack of challenge
- have low self-esteem and a negative attitude towards self and peers
- develop a cynical attitude to hide emotional frustration
- be stifled by an emphasis on reading and writing and lack of creative opportunities
- dominate discussion, with apparently poor listening and turn taking skills
- be socially isolated and rejected by others - though outwardly self-sufficient
- be vulnerable to criticism, responding negatively to activities that reveal the dyslexia
- often feel frustrated, angry, depressed, inadequate
- use humour inappropriately or to attack others
- truant from some subject classes, become disaffected

Aggressive behaviour characteristics:
- refusal to obey classroom rules and display confrontational behaviours
- attention seeking, ignoring the needs of others
- preventing others from engaging with learning, disruptive
- alienation of classmates due to aggression and negative attitude
- appearing bored, frustrated, stubborn and uncooperative
- tactless and impatient with slower thinkers and other learners with dyslexia
- tendency to challenge and question indiscreetly
- masking feelings, sometimes appearing insensitive

Withdrawn behaviour characteristics:
- lacking communication with classmates/teachers
- appearance of time wasting or being preoccupied
- taking a long time to process speech, resulting in slow responses
- preference for working alone at a pace appropriate to the impact of dyslexia
- rarely completing set work in the time allowed
- regarding dyslexia as stupidity, resulting in low self-esteem

Some young people may be confused by the difficulties caused by their dyslexia - even thinking these are due to mental health issues. Some may attempt to conceal their difficulties - even from themselves - by avoiding circumstances where they are likely to fail and by adopting diversionary behaviour. Some young people with dyslexia experience embarrassment, humiliation, anxiety and guilt on a daily basis. When their dyslexia remains unrecognised and unsupported, they lose confidence in themselves as learners, feeling stupid, frustrated and angry.[7]

Teachers might use the Checklists at the back of this booklet to help decide if unacceptable behaviour might conceal underachievement and/or low self-esteem linked to dyslexia.

---

[7] See Case Study - Crossbow's Education Blog - The DIRM Factor – By Beccie Hawes, (Head of Service, Rushall's Inclusion Advisory Team) https://crossboweducation.wordpress.com/2015/03/02/the-dirm-factor/

CLASSROOM MANAGEMENT ISSUES LINKED TO DYSLEXIA

In addition to struggling with literacy issues in class, students with dyslexia may appear:
- to avoid set work
- restless and unable to concentrate
- easily tired, inattentive and uncooperative

It is common for learning differences related to dyslexia that cause unexpected difficulties within the subject curriculum to be mistaken for behavioural issues - students may:
- appear disaffected
- persistently underachieve
- conceal difficulties and will not ask for (or may even reject) help

As a result, teachers may attribute lack of progress to lack of interest or effort, or to misbehaviour. Students with dyslexia often lack stamina and have low self-esteem - having a powerful impact on their ability to cope with the demands of the subject curriculum.

Learners with dyslexia may be slow to respond to a teacher's spoken instructions and find that the rest of the class is getting on with a piece of work while they have no idea where to begin. They may persistently ask for a page number immediately after the teacher has given it. Their inability to remember spoken instructions or process written directions may be interpreted as lack of attention or indiscipline when they ask classmates what to do next.

Many students with dyslexia appear to do everything the long way - concentration is easily lost and they are unable to pick up from where they left off, having to start all over again, leading them to become restless or disruptive to draw attention away from their difficulties. They may start a task well but there is often a rapid deterioration of the quality of work, especially when writing is required. The level of concentration and effort needed for 'normal' class activities may cause fatigue so teachers might vary activities or build-in mini-breaks to allow them to rest briefly.

It is common for students with dyslexia to be disorganised or forgetful of equipment and homework - even turning up in class at the wrong time. Weak organisational skills may be exacerbated by an inability to remember sequences of instructions and a tendency to forget books or to complete tasks on time. Students with dyslexia tend to lose themselves (and their possessions) regularly and they may be unable to retrace their steps because they have no memory of how they arrived at a place.

CLASSROOM MANAGEMENT STRATEGIES

Students with dyslexia are often identified as having behavioural problems when they display:

- short-term (working) memory difficulties
- problems with auditory and/or visual processing
- directional confusion
- weak organisational skills
- poor physical co-ordination

Teachers reporting students' inappropriate classroom behaviour should try to identify the activity that 'triggered' each incident. If 'triggers' can be related to dyslexic difficulties subject teachers might be advised to take account of these when preparing lessons.

Reasonable adjustments:

- allow extra time for dyslexic students to process lesson content - they first have to process the words used, then process the meaning of these
- do not routinely set unfinished class work as homework – dyslexic students take much longer to complete homework than others so there may be a danger of setting too much resulting in incomplete assignments
- accept homework that has been typed or scribed by a parent without question or fuss - though it would be useful to agree this with parents in advance
- check that all tasks are written down correctly and that students with dyslexia understand what is required
- arrange for text material to be read aloud – using a 'reading buddy' or to be converted to digital format in advance so that students with dyslexia can access this
- accept answers in key words/bullet lists or in note form to enable students with dyslexia to get ideas written down - though this skill may have to be taught first
- limit the number of instructions given at one time
- repeat a sequence of instructions at appropriate points during practical activities/ provide a written version of these
- place students with dyslexia near the teacher in order to give individual attention and to encourage them to ask for help (discreetly) when they need it
- ensure that any strategies in the SEN support plan are in place and avoid making comparisons of outcomes achieved with the rest of the class
- mark students' work on content and not spelling/presentation and try to add positive comments - these are rarely experienced by students with dyslexia

Teachers might give some consideration to their method of lesson delivery, which should be multi-sensory and include a variety of activities whenever possible. Secondary subject teachers may be advised to anticipate the possible additional support needs of dyslexic (and other) students when planning lessons and preparing materials.

DEALING WITH INAPPROPRIATE CLASSROOM BEHAVIOUR

When class work that involves a lot of reading and writing is set, students with dyslexia may initially appear to ignore the teacher's instructions. This may be due to an inability to remember spoken instructions or process written directions and not actually linked to inattentiveness or laziness. Students may be checking with classmates what is required, not being deliberately disruptive of the class activity. Teachers might anticipate the impact of set work on students with dyslexia before taking action on any perceived indiscipline.

Barriers to Learning: students with dyslexia may

- seem to make little attempt to settle to the task
- appear to talk to their classmates instead of setting down to work
- be disorganised or forgetful e.g. of equipment, lessons, homework, appointments
- be in the wrong place at the wrong time
- become excessively tired, due to the level of concentration and effort needed for 'normal' class activities

Reasonable Adjustments:

- check whether 'chatting' students are seeking clarification of instructions
- encourage all students to work together and discuss the nature of tasks before starting individual work
- check that all instructions are clear and fully understood - ask students to repeat them aloud
- provide checklists and timetables with schedules and deadlines clearly shown
- vary activities so that students become less fatigued

Bear in mind that most people are more willing to consult the person next to them than to ask the teacher when they are unsure of something. The students being consulted, far from being distracted, often benefit from the opportunity to talk about the work in question, and their own understanding may be clarified and their attention focussed by having the chance to explain something to another student.

MANAGING POOR ORGANISING ABILITY

It will take longer or require more effort for students with dyslexia to achieve the same results as classmates. Teenagers are acutely aware of the gap between their performance and their classmates' apparent ease in some activities. They may not actually realise that this difference is due to dyslexia, especially when this has not been identified, or has been mistaken for behavioural problems.

Barriers to learning: students with dyslexia may

- forget books, work to be handed in, deadlines for assignments etc. and shrug this off, giving the impression that they don't care
- struggle to understand new concepts and be unable to remember sequences of

instructions resulting in accusations that they were not paying attention
- have difficulty taking notes or copying in class - their refusal to admit difficulties with this often results in them being reprimanded for lack of effort
- find it difficult to complete tasks on time leading to them being identified as lazy and not trying
- have difficulty with organisation of homework, so they may refuse to hand it in, preferring to be punished for not doing it rather than admit to problems

Reasonable adjustments:
- provide frequent opportunities for all students to rehearse/practice activities
- provide copies of class notes so that students with dyslexia may listen instead of struggling to keep up with writing
- set tasks that are appropriate to the ability of students with dyslexia but take account of possible barriers to accessing texts or producing written work
- monitor the correct use of the homework diary, put homework tasks online or set them well in advance and involve parents in this
- structure tasks for students with dyslexia, perhaps using flow charts or check lists, **and** help them to prioritise

THE IMPACT OF FATIGUE

The huge effort required by many learners with dyslexia to complete an ordinary task that others can tackle automatically may cause unanticipated fatigue and result in their dyslexic difficulties being attributed to poor attitude or lack of interest. Sometimes students' apparent exhaustion is considered to be due to lack of sleep - and many teenagers will accept this conclusion rather than admit to their daily struggle to keep up in class.

Barriers to learning: students with dyslexia may
- start well but the quality of work quickly deteriorates
- lose concentration easily and become restless or disruptive
- complain of minor ailments or request permission to leave the room
- spend so much time on initial tasks that they do not fully participate in the rest of the lesson
- lack automaticity in ordinary activities
- have to start all over again if interrupted when working
- fall behind the work of the class
- fail to take an accurate note of homework

Reasonable adjustments:
- set short, well-defined tasks with clearly identified outcomes
- vary the types of tasks set and set time limits for their duration
- change activities during the lesson, allowing mini-breaks
- create the opportunity for purposeful movement within the classroom

- teach students how to pace themselves
- give out homework well before the end of the lesson and make sure it is written down correctly – email it home or post it on the school website

SELF-ESTEEM ISSUES

Students with dyslexia often consider themselves to be failures. Some may overreact to remarks made by teachers and peers, taking everything personally. They may be so conscious of the impact of their dyslexic problems that they are oversensitive to casual comments, and respond apparently inappropriately. The stress that is endured by young people with dyslexia in the classroom impacts on their motivation, emotional well-being and even their behavioural stability. Many are working constantly at the limits of their endurance and may rarely - or never - experience success in the classroom.

Barriers to learning: students with dyslexia may
- expect to fail at set tasks, so are reluctant to try anything new
- lack self-confidence
- have a poor self-image
- fear new situations
- be disappointed at a poor return for their efforts
- feel humiliated and embarrassed if their difficulties are exposed in class

Learners with dyslexia may experience despair and exhaustion and be unable to keep up the level of alertness and forward planning needed to sustain intricate coping strategies. When individual coping strategies fail, dyslexia may be expressed by inappropriate diversionary behaviour.

Reasonable adjustments:
- remain aware of students' learning profiles and of the nature of individual strengths and weaknesses
- offer encouragement and support for all activities
- praise effort as well as work well done
- encourage and praise oral contributions
- do not ask students with dyslexia to undertake tasks that might expose them to public failure and humiliation e.g. reading aloud
- mark on **content** not presentation of work

It is important that subject teachers remain alert for difficult situations and defuse any possibly embarrassing circumstances that could result in aggressive or withdrawn behaviours and referral for behaviour support input.

DYSLEXIA - BEHAVIOUR SUPPORT ISSUES

An apparently capable student who appears to be depressed, unmotivated, failing in school and functioning poorly in the family and who may also present violent, disruptive outbursts, will most probably be referred for assessment to the school's behaviour support team (or to the family GP). It is unlikely that any consideration is given to whether dyslexia may be present. By the time an educational psychologist is consulted, any dyslexia indicators may have been ignored or misunderstood and the young person labelled as stubborn, self-centred, unsociable, disruptive or withdrawn and only social/emotional /behavioural interventions are proposed.

While GPs and other health care professionals may not be expected to know a lot about the behavioural, emotional and learning characteristics of learners with dyslexia, it is assumed (often incorrectly) that all teachers and educational psychologists have this knowledge. While some education professionals are fully conversant with the characteristics and intellectual diversity of learners with dyslexia and their typical social, emotional, and behavioural characteristics and needs, classroom teachers are not necessarily aware of these. This lack of information and training of education (and health-care) professionals is the largest single reason for failure to identify or misdiagnosis of dyslexia as a different issue.

Even the physiology of learners with dyslexia may differ from their age-peers - they may have more allergies, sleep problems and uneven rates of development; they may suffer from 'existential depression' -when they feel that they do not fit in the family or the classroom - even wondering whether they are ' mad'.

Dyslexia in able teenagers is often misidentified as a behavioural disorder. The diagnostic process used may be inappropriate - psychological testing, for example, may be carried out in isolation from the context of other sources of information. In order to produce a comprehensive profile of a student, consideration of some of the subtle environmental factors that affect behaviour is essential - including an analysis of the behavioural 'triggers' - e.g. for a refusal to cooperate in the classroom - could quickly indicate the possibility that dyslexic difficulties might be present.

Misidentification of dyslexia as an emotional or behavioural disorder may happen because time constraints lead to snap decisions being made and individuals labelled and treated without adequate understanding of the circumstances that lead to manifestations of the 'problem' behaviour. The SENCo/specialist teacher needs to have time to identify a student's SEN fully before developing plans for support provision – for example:

IDENTIFICATION OF TRIGGERS

Behavioural difficulties in school could be due to a range of factors including:

- learning
- social/emotional
- medical/health
- learning environment
- level of ability
- attendance
- bilingualism

CREATING A LEARNING PROFILE

The SENCo develops a picture of the student based on:

- observations and assessments from school staff (including screening for possible dyslexia)
- information from parents/student
- review of the learning environment
- educational records
- referrals made within school and to other agencies

ACTION TO REMOVE BARRIERS TO LEARNING

Once a student profile has been developed and triggers for behavioural incidents identified, the SENCo - in consultation with the student, parents and other professionals – produces a SEN support plan designed to resolve difficulties – including:

Reasonable adjustments:

- changing the learning environment to be more dyslexia 'friendly'
- introducing strategies to deal with specific/dyslexic difficulties e.g.
    - ❑ individual/small group teaching
    - ❑ peer support/buddying/paired reading
    - ❑ directed in-class support from - teaching assistant, reading buddy
- monitoring attendance at 'problem' subject classes
- setting appropriate target outcomes that take account of any dyslexia
- arranging for staff development input on the behavioural manifestations of dyslexia

Anxiety is a common emotion along with fear, anger, sadness, and happiness, and it has a very important function in relation to the self-esteem of the learner with dyslexia in the classroom. Both voluntary and involuntary behaviours - aggressive or withdrawn - may be directed at escaping or avoiding the source of anxiety - which could be the curriculum, the learning environment or even the subject teacher.

FURTHER READING

DES (Eire) *Dyslexia and Challenging Behaviour* Special Education Support Service, Cork
http://www.sess.ie/dyslexia-section/dyslexia-and-challenging-behaviour

MacKay, N (2005) *Removing Dyslexia as a Barrier to Achievement: The Dyslexia Friendly Schools Toolkit* 3rd Edition (2012) Wakefield, SEN Marketing

Pavey, B (2012) *The Dyslexia-Friendly Teacher's Toolkit* London, Sage
This book is a really practical, hands-on guide packed with a wealth of advice on strategies and "things to try" reflecting the author's extensive experience.

Peer, L & Reid, G (2001) *Dyslexia – Successful Inclusion in the Secondary School* London, David Fulton Publishers Chs. 1, 26 & 30

Reid, G & Fawcett, A (Eds) (2004) *Dyslexia in Context - Research, Policy and Practice* London, Whurr Chs. 7, 12 &19

Reid, G (2009) *Dyslexia: A Practitioner's Handbook* 4th Edition Chichester, Wiley-Blackwell Chs. 14-16

Reid, G (2013) *Dyslexia and Inclusion Classroom approaches for assessment, teaching and learning* (2nd edition) Abingdon, Routledge
Now fully updated, this book aims to equip all teachers with the necessary knowledge of dyslexia in order to for it to be effectively understood and dealt with in the classroom.

Rooke, M (2015) *Creative Successful Dyslexic* London, Jessica Kingsley Publishers
Well-known people talk about how dyslexia affected their childhoods and how they overcame the challenges and used the special strengths of dyslexia to achieve great success in adulthood.

Schultz, J (2013) *Dyslexia Stress-Anxiety Connection* International Dyslexia Association
http://eida.org/the-dyslexia-stress-anxiety-connection/

Thomson, M (2008) *Supporting Students with Dyslexia at Secondary School: every class teacher's guide to removing barriers and raising attainment* Abingdon, Routledge Ch. 1

Thomson, M (2008) *Dyslexia and Inclusion at Secondary School* IN Reid et al (Ed) 2008 *The SAGE Handbook of Dyslexia* London, Sage

West, T G (1997) *In the Mind's Eye: Visual Thinkers, Gifted People with Learning Difficulties, Computer Images and the Ironies of Creativity* Loughton, Prometheus

DYSLEXIA INDICATORS AT THE SECONDARY STAGE (PHOTOCOPIABLE)

Dyslexia is more than an isolated defect in reading or spelling. The problem may be perceptual, auditory receptive, memory-based or a processing deficit.

Subject teachers are not expected to be able to diagnose these difficulties as such, but some general indications are listed below. If several of these are observed frequently in the classroom, please tick the relevant boxes and enter details- of the student concerned and pass to the SENCo/specialist dyslexia teacher for information/investigation.

Student Name: _____ Class: _____ Date: _____

- ☐ Quality of written work does not adequately reflect the known ability of the student in the subject
- ☐ Good orally but very little written work is produced - many incomplete assignments
- ☐ Disappointing performance in timed tests and other assessments
- ☐ Poor presentation of work - e.g. illegibility, mixed upper and lower case, unequal spacing, copying errors, misaligned columns (especially in Maths)
- ☐ Weak organisational skills - student is unable to organise self or work efficiently; carries either all books or wrong ones; frequently forgets to hand in work
- ☐ Sequencing inconsistent - student appears to jump from one theme to another, apparently for no reason
- ☐ Inability to memorise (especially in Maths and Modern Languages) even after repeated practice
- ☐ Inability to hold numbers in short-term memory while performing calculations
- ☐ Symbol and shape confusion (especially in Maths)
- ☐ Complains of headaches when reading; sometimes sees patterns in printed text; says that words move around the page or that text is glaring at them
- ☐ Unable to carry out operations one day which were previously done adequately
- ☐ Unable to take in and carry out more than one instruction at a time
- ☐ Poor depth perception - e.g. clumsy and uncoordinated, bumps into things, difficulty judging distance, catching balls, etc.
- ☐ Poor self-image - lacking in confidence, fear of new situations - may erase large quantities of written work, which is acceptable to the teacher

- ❑ Tires quickly and work seems to be a disproportionate return for the effort involved in producing it
- ❑ Easily distracted - either hyperactive or daydreaming
- ❑ other - enter details

Teacher_____ Subject: _____

Action/information requested:

- ❑ details of known SEND and support required
- ❑ investigation of SEND and advice on graduated support
- ❑ dyslexia screening/assessment
- ❑ profile of learning needs
- ❑ suggest reasonable adjustments to be made in class
- ❑ suggest learning objectives and outcomes for SEN plan
- ❑ advice re Access arrangements

DYSLEXIA: CHECKLIST OF UNDERACHIEVING BEHAVIOUR (PHOTOCOPIABLE)

Subject teacher: _____

Student details: _____

Learning characteristics

- ❑ Is orally good but written work is poor - gap between expected and actual performance - may be reluctant to write at length because s/he cannot write as fast as s/he thinks *how well it*
- ❑ Is apparently bored, appears to be absorbed in a private world
- ❑ Often abandons set work before finishing, having mastered content/process
- ❑ Can follow complex instructions easily, but prefers to do things differently
- ❑ Works independently, but finds many reference sources superficial
- ❑ Good problem finding skills, but reluctant to solve these once identified
- ❑ Inventive in response to open ended questions, able to form but not test hypotheses
- ❑ At ease in dealing with abstract ideas
- ❑ Shows a vivid imagination with unusual ideas
- ❑ Is very observant, sometimes argumentative, able to ask provocative questions

Behavioural characteristics

- ❑ Has a poor concentration span but is creative and persevering when motivated
- ❑ Seems emotionally unstable - may have feelings of inferiority but is outwardly self-sufficient
- ❑ Often restless and inattentive, lacks task commitment
- ❑ Prefers to work alone, rarely co-operates in group work
- ❑ Shows originality and creativity but is quickly bored with repetitive tasks
- ❑ Has a narrow range of interests and hobbies with extraordinary knowledge of obscure facts
- ❑ Appears to have little in common with classmates, being tactless and impatient with slower minds
- ❑ Has a quirky, sometimes adult, sense of humour

DYSLEXIA - SELF-ESTEEM ISSUES (PHOTOCOPIABLE)

The student with dyslexia needs a great deal of support and encouragement to help face up to, talk about and analyse those confusing and conflicting emotions and behaviours that can result from what is often called the 'hidden disability'

Learners with dyslexia, unless gifted and talented in a specific area, may go through the school system never knowing the experience of success.

Student Name: _____ Class: _____ Date: _____

Please indicate any of the following that you suspect this student may be experiencing:

- ❑ lack of self-confidence
- ❑ poor self-image
- ❑ a fear of new situations
- ❑ fatigue from the huge effort needed to complete an ordinary task that others can tackle automatically
- ❑ disappointment at the disproportionate return for their effort
- ❑ confusion regarding their place in the 'pecking order' of the class, which often leads to isolation or identity problems
- ❑ humiliation as their difficulties lead to embarrassing situations
- ❑ despair and exhaustion from the effort to maintain the alertness and forward planning needed to sustain intricate coping strategies

Please note any other issues you have observed re this student and return to: _____

```


```

Signed:_____